Welsh Railway Records
Volume 2:
Barry Railway Drawings

The original company crest was a circular green garter device with the company name in the garter and "B R" in the centre. This was carried on the locomotives, either on the cab side or on the side tank below the cab panel. The Belgian built engines of 1899 had them on the bunker sides, their builder's plates being on the tank sides. Before leaving the Company in 1905, Mr Hosgood introduced a new crest, shown here, of an oval green garter with Company title and date of incorporation, 1884, the white centre bearing a red dragon and a stag's head full face at top centre. The latter was taken from the Coat of Arms of the Windsor–Clive family, as Lord Windsor, who later became the Earl of Plymouth, was the Company's only Chairman from 1884 to 1922.

This page: Class H No. 79 stands resplendent in its builder's photographic grey livery in September 1896. Together with the Class D, the Class H was the most powerful locomotive type employed by the Barry Railway. They were used to take coal trains from the marshalling yard in Cadoxton to the sidings above Barry Docks, known as Gladstone Sidings, from where the wagons were shunted to their respective ships. Because of their power, they could haul coal trains of over 100 wagons. SHARP, STEWART & CO

Front cover: Class J No. 98 stands in platform 3 at Barry station waiting to depart for Cardiff Clarence Road. The change in numbering from brass numbers on the front of the chimney to gilt lettering on the front buffer beam took place after 1905 by order of Mr Golding, Locomotive Superintendent from that year until 1909. In the background to the left of the engine can be seen Barry carriage shed. To the right of the engine at the rear of the platform, the east end of Barry Junction Signal Box is visible behind the passengers boarding the train. PHOTOGRAPHER UNKNOWN

Back cover: Barry Docks offices. This Baroque Revival building was constructed between 1897 and 1900 though the date on the building is given as 1898 and it was designed by the Cardiff-based architect Arthur E Bell. It was built at about the same time as Dock No. 2 by which time Barry had established an international reputation as a coal-exporting port and the Company required a building to go with its new-found status. The building comprises red brick and Portland stone and the central doorway with its concave surround is said to be modelled on a Wren design for one of his London churches. The bronze statue to the right, designed by Alfred Gilbert of London, is of David Davies, one of the main protagonists behind the formation of the Barry Dock and Railway Company; a copy of this statue is to be found at his birthplace in Llandinam, Powys, which was also where he died. PHOTOGRAPHER UNKNOWN

Welsh Railway Records
Volume 2:
Barry Railway Drawings

Trefor Jones, Mike Morton Lloyd
and Rhobat Bryn Jones

Welsh Railways Research Circle

Rhagair

Yr Ail Gyfrol yw hon o Gyfres Cofnod Rheilffyrdd Cymru sydd yn cynnwys arluniau Trefor L Jones a Mike Morton Lloyd a bortreada locomotifau Rheilffordd y Barri yn ogystal â cherbydau a wageni. Yn wreiddiol, cyhoeddwyd y lluniau hyn yn y wasg fodelu Rheilffyrdd o'r 1940au ymlaen. Cynhwysir arluniau gyda'r manylion angenrheidiol i gyd â ffotograffau, er mwyn cynhyrchu modelau manwl gywir.

Pan fu farw Mike, rhoddodd ei weddw, Irene, ganiatâd i Gylch Ymchwil Rheilffyrdd Cymru gael defnydd o'r arluniau gwreiddiol. Calon a diben y llyfr hwn a'r gyfres ydynt. Felly, hoffem fynegi a chofnodi ein diolchgarwch i weddwon Trefor a Mike, sef Betty ac Irene, am eu haelioni gan ganiatáu i'r Cylch gyhoeddi'r lluniau gwerthfawr hyn.

Hoffem ddiolch i Richard Evans hefyd a gyfranodd i waith y cyhoeddiad hwn gan gwblhau lluniau Dosbarth C Rheilffordd y Barri, fersiwn y 2-4-0T a'r 2-4-2T ill dau yn ogystal â'r Car Modur.

Adnodd amhrisiadwy yw'r arluniau hyn, nid yn unig ar gyfer modelwyr ond ar gyfer haneswyr hefyd sydd arnynt eisiau dysgu mwy am y cyfnod hynod ddiddorol hwnnw rhwng 1836 a 1922 pan oedd gan Gymru gymaint o'i chwmnïau rheilffyrdd ei hun. Gobaith Gweithgor Cyhoeddiadau'r Cylch yw i chi gael y llyfr hwn yn llawn gwybodaeth o ddiddordeb ac yn rhoi llawer o oriau o ddarllen difyr.

Rhobat Bryn Jones
Mis Rhagfyr 2018

Foreword

This is Volume 2 in the Welsh Railway Records series, which includes the drawings of Trefor L Jones and Mike Morton Lloyd, depicting Barry Railway locomotives and other rolling stock. These drawings were originally published in the railway modelling press from the 1940s onwards. They include all the necessary details, together with photographs, for accurate models to be produced.

On Mike's death, his widow, Irene, kindly allowed the Welsh Railways Research Circle access to the original drawings and they form both the heart and purpose of this book and the series. We wish, therefore, to express and record our gratitude to the widows of Trefor and Mike, Betty and Irene, for their generosity in allowing the Circle to publish these valuable drawings.

Thanks are also due to Richard Evans who has contributed to the work in this publication by completing drawings of both the 2-4-0T and 2-4-2T versions of the Barry Railway Class C as well as the Motor Car.

These drawings are an invaluable resource not just for modellers but also for historians who wish to learn more about that fascinating period from 1836 to 1922 when Wales had so many of its own railway companies. We, the Publications Working Group of the Circle, hope you find the book informative and provides you with many hours of enjoyable reading.

Rhobat Bryn Jones
December 2018

WELSH RAILWAY RECORDS

Drawings by
Original drawings by Trefor L Jones and Mike E M Lloyd with new additions by Richard Evans

Designed by Jonathan David

ISBN: 978 0 9527267 8 4

Published by the
Welsh Railways Research Circle
Rhiwderin, Newport NP10 8RL
www.wrrc.org.uk

Printed by Short Run Press Ltd, Exeter

Contents

Setting the scene 7
Liveries 11

Locomotives
BRL1 0-6-0T Class A Locomotives 16
BRL2 0-6-2T Class B Locomotives 20
BRL3 0-6-2T Class B1 Locomotives 24
BRL4 0-6-2T Class B1Locomotives as rebuilt by the GWR 28
BRL5 2-4-0T Class C Locomotives 30
BRL6 2-4-2T Class C Locomotives as rebuilt from the 2-4-0T by the Barry Railway 32
BRL7 2-4-2T Class C Locomotives as rebuilt by the GWR 34
BRL8 0-8-0 Class D Locomotives 38
BRL9 0-6-0T Class E Locomotives 42
BRL10 0-6-0T Class E Locomotives as rebuilt by the GWR 44
BRL11 0-6-0ST Class F Locomotives 46
BRL12 0-6-0ST Class F Locomotives as rebuilt by the GWR 49
BRL13 0-4-4T Class G Locomotives 52
BRL14 0-8-2T Class H Locomotives 56
BRL15 2-4-2T Class J Locomotives 60
BRL16 2-4-2T Class J Locomotives as rebuilt by the GWR 62
BRL17 0-6-2T Class K Locomotives 66
BRL18 0-6-2T Class K Locomotives as rebuilt by the GWR 68
BRL19 0-6-4T Class L Locomotives 72

Motor Cars
BRM1 Motor Cars 78
BRM2 Motor Cars after Conversion to Hauled Stock 82

Carriages
BRC1 Third Class Carriages Nos 27–41 86
BRC2 First/Second Composite Carriages 158–174 88
BRC3 Brake/Third Carriages Nos 197–204 Originally numbered 1–8 90
BRC4 Brake/Third Carriages Nos 205–208 Originally numbered 9–12 92
BRC5 Third Class Carriages Nos 2, 3, 4, 6, 8 and 10 (new series) 94
BRC6 Tricomposite Carriages Nos 150, 151, 152 and 156 (new series) 96
BRC7 Brake Third Carriages Nos 197–200 (new series) 98
BRC8 Bogie used on 1920 carriages 101
BRC9 Details for 8 ft 6 in. bogie used on 1920 stock 102
BRC10 Detail of Guard's Compartment, 1920 stock 103
BRC11 End elevation and cross-sections of First and Second Class Compartments for 1920 carriage stock built by the Birmingham Railway Carriage & Wagon Company 104
BRC12 Details of Interiors of Compartments for 1920 Carriage Stock built by the Birmingham Railway Carriage & Wagon Company 106
BRC13 Detail of carriage side, Tricomposite, 1920 stock 108
BRC14 Detail of carriage side and longitudinal section through Third Class compartment, 1920 stock 109

Mineral Brake Vans
BRB1 Four-wheeled Brake Vans Nos 1–61 110
BRB2 Six-wheeled Brake Vans Nos 62–73 112

Goods Vehicles
BRW1 Five Plank 10 Ton Open Goods Wagon with Tarpaulin Rail 114
BRW2 Four Plank 10 Ton Open Goods Wagon 115
BRW3 10 Ton Timber Set 116
BRW4 10 Ton Covered Van 117
BRW5 10 Ton Cattle Truck 118
BRW6 10 Ton Iron Bodied Van 119

Numerical Listings
Locomotives 120
Motor Cars and Conversions 125
Passenger Rolling Stock 126
Mineral Brake Vans 132
Goods Traffic Rolling Stock at the Grouping
Revenue Stock 133
Ballast Wagons 140
Coal Wagons 142
Miscellaneous and Duplicate Numbered Wagons 143

Bibliography and Further Reading 144

The chief advantage that the Barry Railway had over its main rival, the Taff Vale, was that it was an integrated dock and railway company as this photograph well illustrates. At its centre is the Barry Dock Offices, the fulcrum of the entire operation. This photograph was taken during the early days of the Grouping. At the top is No. 2 Dock with the Rank Hovis Atlantic Mills showing top left, built in 1904, next to the transit shed to its right. To the left of the Dock Offices can be seen the white end of the Scandinavian Church and to the right, two graving docks. The one nearest the offices was built in 1889 and the one further right was built in 1893.

Setting the Scene

An Introduction to the Barry Railway

The Birth of the Railway

The Barry Railway owed its existence to market failure, the failure of the market to match the demand for transporting coal with the supply of adequate facilities for it to reach its customers worldwide. The Rhondda colliery owners were producing increasing amounts of coal each year but, because of their valley-side locations, there was precious little room for sidings that could store coal wagons until they were ready for dispatch. The coal owners had to rely on two elements of transport infrastructure in order to reach their markets: the Taff Vale Railway and Cardiff Docks, the latter managed by the Bute Trustees. In effect, the Trustees had a monopoly on coal exports from the South-East Wales region, though it faced competition from Newport, and the Taff Vale had a monopoly on transporting coal from the Rhondda to Bute Docks.

There had been previous attempts to construct a dock and railway at Barry, in 1865 and in 1877, though it was the increasingly dysfunctional infrastructure that brought matters to a head. Sustained lobbying by the coal owners of the Bute Estate for an expansion of Cardiff Docks continued for many years. Finally, in 1882, a Bill was brought forward to revive the building of Roath Dock; but it came at a price. To finance it, a penny would be added to each ton shipped, not just from Roath Dock but from all of the docks under the Trustees' control. This perceived abuse of their market position was the spark that re-ignited the Barry Dock project.

The first Bill to promote the scheme was defeated by the vested interests who had nothing to gain from a change in the *status quo*. In addition to the Bute Trustees and the Taff Vale could be added the Rhymney Railway which, enjoying favoured status at Cardiff Docks, no doubt felt it politic to side with the Trustees and cement the good commercial relations that had existed between the two businesses.

Undeterred, the subscribers presented another Bill. The financial costs of an increasingly malfunctioning transport network continued to worsen. This second Bill received Royal Assent on 14th August 1884 as the Barry Dock and Railways Act and the Barry Dock and Railway Company was born.

Barry Dock

The tender for the construction of the Dock, which cost £2 million, was awarded to T A Walker, who was also in the process of constructing the Severn Tunnel and the Manchester Ship Canal at the time. Lord Windsor, Chairman of the Company, had the honour of cutting the first sod on 14th November, a Friday, in 1884. The area to be contained by the Dock was in the region of 200 acres. By a process of trial and error, the contractor finally succeeded in excluding tidal waters from the Dock in March 1886.

However, this was only the beginning. The work of creating the dock now had to proceed. This involved the excavation of 5 million cubic yards of soil, the laying of 200,000 cubic yards of masonry and 20,600 cubic yards of brickwork and the use of 175,000 cubic feet of timber. At its height, the work was carried out around the clock and some 3,000 men and 31 locomotives were required to complete the task. The stone used in the construction was obtained from two different sources, both local: the shores of Sully and Alps Quarry in Wenvoe.

When completed, the Dock area was 73 acres in total, 3,400 ft long by 1,100 ft wide. From the perspective of competing for local trade, the biggest dock in Cardiff was Roath Dock at 46½ acres. A comparison with the Albert Dock in London, which covered 72 acres, shows that the significance of this development went beyond the South Wales coalfield region. Initially, Class A Nos. 1 to 4 were the main shunting locomotives for the docks, but in 1890 these began to be replaced by the Class F saddle tank which, by 1905, totalled 28 engines. The Class E was introduced

Dinas Powis station. Barry Railway Class J No. 98 pulls out of Dinas Powis station hauling a rake of ex-Lancashire & Yorkshire coaching stock on its way to Barry with a McKenzie & Holland somersault signal showing the road. The roof of the signal box can be seen at the bottom of the picture. J PITT

between 1889 and 1891, to negotiate some of the sharper curves to be found in the docks.

Perhaps the amenity that most characterised the modern nature of the Dock was the coal-tip. It was essentially a tower, the top of which could be reached by rail from the high-level lines that radiated out from the Gladstone sidings area just to the north of the dockside. There were 17 of them in total and, with the advantage of moveable staiths, several of a ship's hatchways could be loaded simultaneously. It was originally calculated that with an average of 14 coal tips working at any one time, each one could load 4,300 tons a week which meant that over the year, the total would be 3,130,000 tons.

The Main Line

The main line was built specifically to overcome the congestion on both the Taff Vale and at Cardiff Docks. From a junction with the Taff Vale Railway at Trehafod, the Barry company's route began its 19 mile journey to the coast. Between Trehafod and Pontypridd stations were the Exchange Sidings at Hafod; this is where Rhondda coal traffic transferred to Barry company rails. After Hafod, the line turned south through Pontypridd where it entered a tunnel, 1,323 yards in length, before emerging at Treforest. The line continued south through Efail Isaf, Creigiau, crossing the St Fagans Viaduct (180 yards) over the Great Western Railway and the Drope Viaduct (178 yards) over the River Ely, into the 1,867 yard Wenvoe Tunnel before emerging into daylight at Wenvoe. The line continued south before passing through Cadoxton Yard and into Cadoxton, the junction station where the main line and the Cogan Branch diverged.

It was on this section of the line that the greatest variety of locomotive stock was to be found. The main mineral traffic was hauled by both Class B and B1 locomotives. In 1899, the Class K was introduced to supplement the use of the Class B1 in hauling mineral trains but they were found to be costly in terms of coal and water usage. As a result, they were re-assigned to Trehafod and Treforest on banking duties and to assist in shunting duties between Cadoxton Yard and the Dock. Observers, from 1905 onwards, would also have been able to see the Class D hauling mineral trains from the Rhymney Valley.

Passenger services came from two directions: heading north from Barry, which started in March 1896; and from Cardiff Clarence Road, which started in June 1897. In both cases, Class G locomotives hauled the passenger trains with the exception of the workmen's trains which were hauled by the Class B1.

Cadoxton Yard was the effective terminus of coal traffic from Rhondda and from the Cynon and Taff Valleys via the exchange sidings at Treforest. At its most extensive in 1914, the Yard had 18 Up sidings (for Rhondda and beyond) and 15 Down sidings (for the Dock). It also, from 1903 onwards, had an overhead signal box towards the north of the yard that lasted until the Barry's amalgamation with the GWR in 1922. It was here that coal trains would terminate and the trucks be marshalled into formations appropriate for the loading of ships. These trucks would then be taken, often in trains of 100 wagons, to the sidings above the Dock, known as Gladstone Sidings. This traffic was initially hauled by the Class D and then, from 1896 onwards, by Class H locomotives of which there were seven. They were later joined by the Class K after the failure of this class to haul mineral trains on the main line. In Gladstone Sidings, trains could be sub-divided again and allocated to the various coal tips that lined the dockside.

The size of the Dock estate can be best imagined by realising that Cadoxton station marked the eastern boundary of Barry Dock whereas Barry station, marked its western boundary, stations that were two miles apart.

The Freight Branches

From its inception, the business strategy of the Barry company was not to serve collieries directly but rather to siphon off coal traffic from other companies and redirect it to Barry Dock. To this end, the 1884 Act included provision for connecting lines between Treforest on the Taff Vale and Tonteg on the Barry company's line; from Peterston on the GWR main line to Drope on the Barry company's main line; and also from St Fagans on the GWR main line to Tynycaeau Junction to the north of St Fagans Viaduct.

A later branch, discussed below, was to Cardiff and this made possible the transfer of coal traffic from the Gwent Valleys at Penarth Curve, just to the west of Cardiff General station. However, transferring coal traffic so close to a main-line station and motive depot proved difficult. Therefore in later years, the Barry company was given permission to haul trains from Rogerstone; the Class L was often assigned to this duty, along with the Class B1.

The Passenger Branches

The first scheduled trains to run on the Barry Railway were not on the main line but on its first official branch line from Cadoxton to Cogan known, not unsurprisingly, as the Cogan branch. This required the construction of a tunnel between Dinas Powys and Cogan, some 222 yards in length. Although the first locomotive to run on the system was on the main line on 22nd November 1888, the first scheduled passenger service ran on 20th December 1888, from Cogan to Barry Dock with intermediate stations at Dinas Powys and Cadoxton, Barry station becoming the new terminus for the service on 8th February 1889; it was hauled by Class A No. 5. This single locomotive was replaced soon afterwards by the Class C in the same year. Four locomotives of this class handled the passenger traffic on this five-mile journey. The service to Cogan would later be extended to Cardiff Riverside on 14th March 1893 (whereupon the Cogan branch became the Cardiff branch) and to Cardiff Clarence Road on 2nd April 1894, thus lengthening the journey to eight and ten miles, respectively.

1892 saw the introduction of the Class G which replaced the Class C on the Cardiff branch. Five years later, between 1897 and 1899, the Class J was delivered which, in turn, replaced the Class G in hauling the Barry's busiest suburban passenger service.

The second passenger branch to be built was from Barry to Barry Island which opened on 3rd August 1896 and added a mile to the network. This was later extended to Barry Pier on 27th June 1899 by means of a 280-yard tunnel running east–west. This branch was unusual in that it did not follow the strategy of trying to siphon other companies' minerals business but rather sought to tap into a new market: leisure. Barry Island, when it was an island, had always been a favourite destination for tourists but on a very small scale. Now that the island was connected to the mainland not only by a causeway but by a modern railway, the popularity of the destination grew at a phenomenal rate. As an example of this, a switchback railway that had been installed for the Cardiff Empire Exhibition in 1896 was moved to Barry Island for the benefit of the new influx of tourists arriving by train. In the years to come, Barry Island would be the destination for many an excursion train from all parts of the UK, usually hauled by a Class B1.

Barry Pier was an unusual station in that it did not have a regular scheduled service but rather accommodated trains to meet the steamers that berthed at the pier, which was initially built for the use of P & A Campbell steamers. The station was adjacent to a floating pontoon dock so that travellers would have ease of transfer from one form of transport to another. The station had

Cogan station. In the foreground is the Barry Railway station at Cogan with Penarth Docks in the background. Penarth Dock and Harbour station can be seen between them. Until 1893, Barry trains terminated at Cogan so that passengers for Cardiff had to cross to the Taff Vale station at Penarth Dock. Complaints were received at the time that during bad weather the crossing was particularly unpleasant because of the rain and mud underfoot. FROM A CONTEMPORARY POSTCARD

an important function which was to take passengers to the Barry's own fleet of paddle-steamers known as the Red Funnel Line, a service which began in 1905. The company was nothing if not enterprising in its discernment of new business opportunities for a line whose initial purpose was to transfer coal.

The Vale of Glamorgan Railway

Another opportunity to expand its connections into other companies' coal markets came with the arrival of the Vale of Glamorgan Railway in Barry on 1st December 1897. This had begun as an independent company but had had difficulty in raising the necessary finance. The Barry company stepped in with an offer to run the railway and guarantee a 4% dividend on the shares in exchange for 66% of the Vale of Glamorgan receipts. Construction of the 19-mile line involved a tunnel (No. 1 – 543 yards), a viaduct (376 yards) and a second tunnel (No. 2 – 73 yards) at Porthkerry. The line, which was double-tracked throughout, ran from Barry to Bridgend with intermediate stations at Rhoose, Gileston, Aberthaw, Llantwit Major and Southerndown Road.

From the Barry company's point of view, this line would provide access to the coal traffic emanating from the Llynfi, Garw and Ogmore Valley collieries which, up to that point, had had to use the somewhat deficient dockside facilities at Porthcawl. Although the strategy had worked well elsewhere, two factors served to conspire against the same success being obtained here. First the docks at Port Talbot were substantially enhanced after the passing of the Port Talbot Railway & Docks Act 1894, and this proved to be a more convenient outlet for that particular traffic. Secondly, the new junction at Coity near Bridgend, which the Barry hoped would divert coal traffic from the GWR, was in fact bypassed by that company. It made more commercial sense for its traffic to travel on to the junction at Peterston and thus gain additional revenue for the extra mileage covered.

Mineral and goods traffic were largely hauled by Class B1 locomotives whereas the passenger traffic saw a number of different train formations in its time. Regular passenger traffic was hauled by the Class C and Class J, but for a short while, between 1905 and 1909, steam motor cars were used to supplement services between Barry and Llantwit Major after their failure on the Cardiff–Pontypridd service. Subsequently, they were replaced by a Class E locomotive (normally No. 33) coupled to a six-wheel Brake Third and a four-wheel Composite, known as a "Motor Set", of which there were two. Ironically, the Motor Cars returned to the line in 1914 but as trailer coaches only, their power units having been removed. They were coupled to a Class C locomotive and became known as the "Vestibule Set".

Eastward to the Rhymney Valley

The last line to be built by the Barry Railway was the only part of the system that did not carry passengers on scheduled services; it was a freight line only and was known as the Penrhos Branch. It left the main line at Tynycaeau Junction and headed east through Walnut Tree Tunnel, across Walnut Tree Viaduct, towards Caerphilly where it met the Rhymney Railway at Penrhos Lower Junction. This part of the branch was opened on 1st August 1901. The exchange sidings here enabled the Barry company to divert coal traffic, once again, towards Barry Dock. However, this was not the extent of the Barry's ambition as it opened the extension to the Brecon & Merthyr Railway on 2nd January 1905. After Penrhos Junction, the line turned north across another two viaducts, Pen-yr-Heol and Pwll-y-Pant, before joining the Brecon & Merthyr Railway just south of Maesycwmmer at Barry Junction, a distance from Tynycaeau of just over nine miles. The journey over the B&M rails to its terminus was an additional 14 miles.

In short, no expense was spared to gain direct access to the collieries on the eastern side of the Rhymney as the Barry had running powers as far as the terminus of the line, Rhymney & Pontlottyn. Both Class B1 and Class D locomotives could be seen hauling the mineral traffic on this branch.

Commercial Success

It is easy to oversimplify, and thus misrepresent, the financial performance of any company over a 23-year period, but there are certain key figures that indicate how successful this company was. In 1889, the coal and coke tonnage exported from Barry Dock was 1,073,575. By 1898, that figure had increased to 4,369,448

Efail Isaf station was to be found on the Barry Railway main line which opened on 8th July 1889. The company was reluctant to provide passenger services and residents along the line submitted a petition on 16th May 1890. However it was not until 16th March 1896 when passenger trains were provided between Barry and Porth. R M CASSERLEYY 19062, 15 JULY 1959

tons. Based on the favourable trading figures, the decision was taken by way of a Parliamentary Act in 1894 to open a second dock, Barry No. 2 Dock, in 1898.

The following year, in 1899, coal and coke exports rose to 7,223,669 tons, an increase of 65% on the previous year, and in 1901 Barry Dock overtook the Bute Docks for the first time in coal tonnage exported. It reached its zenith in 1913 with a figure of 11,049,711 tons. Demand for space at the Dock continued to rise and the Barry company did not make the same mistake as the Bute Docks had previously. Six new coal tips were erected at Barry No. 1 Dock, this time on the south side. In railway terms, 150 miles of sidings were required to deal with the coal traffic, 100 of those being in the dock areas alone.

This success was reflected in the financial performance of the company. Between 1890 and 1897, dividends on ordinary stocks and shares, which were paid half-yearly, dipped below 10% on only three occasions, in December of 1891, 1892 and 1893. The dividend paid at the end of each those half-years was 9%. Subsequently, the Company's business was affected by various events outside of the company's control, such as the colliers' strike between April and September 1898. This resulted in no dividend being paid for the half-year ending June 1898. Similarly, a six-month strike at the Cambrian Collieries in 1911 resulted in a reduced dividend of 6% in June 1911.

Conversely, the Barry company was able to benefit from the misfortunes of other companies. The Taff Vale strike of 1900 led to a substantial increase in the Barry's income. Later, however, the Barry company rarely reached the height of its financial performance that it had achieved in its early years. Between 1898 and 1913, a dividend of 10% was paid out for only four half-year periods. The last of these was in 1913 when its record-breaking export figure was achieved. After this, because of the First World War, railways came under Government control with a guarantee of net receipts being provided. It would not be long after the war before the Barry company would be amalgamated with the Great Western Railway at the Grouping.

But, as Dylan Thomas was to write some years later, they did not "go gentle into that good night". It is fitting to quote the Earl of Plymouth who said at the last ordinary yearly meeting on 17th February 1922:

"The Barry Railway Company was authorised by Parliament in the face of strenuous opposition, and its development was made in the same spirit, but notwithstanding this the Company's progress has been one uninterrupted success from its commencement, and your Directors have every confidence that if it were possible for the Company to have remained as an independent undertaking, they might have anticipated a period of still greater prosperity."

Barry Railway boilers

From the beginning, the Barry Railway was intent on using standardised boilers for all its locomotives. There were in essence three types of boiler which were later categorised as Types 1, 2 and 3.*

Type 1 was the original two-ring boiler with 150 lb/in.^{2} of pressure as fitted to the Class A and were also used on the first 15 deliveries (Nos. 6 to 20) of the Class B. This boiler was never modified. The Type 2 was a three-ring boiler, again with a pressure of 150 lb/in^2. This boiler was used on the Class B (Nos. 23 to 32), and the Classes B1, F, G and J. The Type 3 was also a three-ring boiler with 160 lb/in^2 pressure.

The tube arrangements of Types 2 and 3 were modified during their lifetime which led to a further sub-categorisation of A, B and C. For example the Type 2 boiler arrangements were as follows:

Type 2A: 181 × 2 in.
Type 2B: 182 × 1⅞ in.
Type 2C: 184 × 1⅞ in.

Prior to 1909, the official Barry Railway boiler numbering system had been based on the locomotive number with a suffix added to denote a replacement. For example, locomotive No. 38 would have been fitted with boiler No. 38. When its replacement was fitted, this would have been numbered 38A, the second replacement 38B and so on. Following this date, although boilers continued to be numbered, the systematic approach broke down and the numbering of the boilers no longer corresponded with the locomotive's number.

* *The categorisation is to be found in Part Ten (Absorbed Engines) of "Locomotives of the Great Western Railway" published by The Railway Correspondence and Travel Society in 1966. It was not used by the Barry Railway.*

Liveries

By Cliff Harris, reprinted with minor amendments from "Great Western Way", second edition, by kind permission of Cliff Harris and the HMRS

The history of the Barry Railway livery divides into three phases, based on the three Locomotive Superintendents, Mr J R Hosgood 1888–1905, under whom nearly all Barry locomotives and stock were built; Mr H F Golding 1905–1909; and Mr J Auld 1910 to the Grouping.

Under Mr Hosgood, locomotive livery was described as brown and all classes were lined. The original style was a scarlet band around the edges of the panels; within and separate from it was lining of a narrow black band edged either side by thin yellow lines. Up to 1891, when the Company's title changed from Barry Dock & Railway Co to Barry Railway Co, the oval numberplates were lettered "BARRY" above the number.

With the new title in 1892 the plates changed to "BARRY" above the number and "RAILWAY" below. All the locomotives built up to 1905 carried this pattern. The background colour of all patterns used was black.

The original company crest was a circular green garter device with the company name in the garter and "B R" in the centre. This was carried on the locomotives, either on the cab side or on the side tank below the cab panel. The Belgian built engines of 1899 had them on the bunker sides, their builders' plates being on the tank sides.

Before he left, Mr Hosgood introduced a new crest of an oval green garter with company title and date of incorporation, 1884, the white centre bearing a red dragon and a stag's head full face at top centre. This crest appeared on the steam motor cars of 1905 and the locomotives built that year. After the side chains originally fitted were removed from the buffer beams they were lettered "B R" in gold shaded blue lettering.

The first Barry coaches were 36 four-wheelers built between 1888 and 1890. The livery was crimson lake sides, ends and solebars, lined with thin chrome yellow lines either side of the mouldings. Lettering and numbers were in gold shaded to the right and below in mid-blue, in plain block letters without serifs. Class identity was by "FIRST", "SECOND", "THIRD" in door waist panels as normal. The brake vans had "GUARD" and "LUGGAGE COMPT" and originally had one Third-class compartment at one end (later built into the van space).

Six-wheel coaches were introduced in 1893 with a batch of six Thirds; more Third Composites and Brake Vans following from 1895 to 1900. The Thirds and Composites followed the style of the four-wheelers, lengthened, but the Brake Vans now had the Guard and van in the centre and a Third-class compartment at either end. When new they carried the circular crests and the number and "B R" twice in the waist panels. The Brake Vans only had the crests at one end, the dog box door being in that spot at the other.

The four-wheelers were oil lit but electric lighting appeared from 1897 onward in the six-wheelers. In 1902 the Manager sought permission to have the four-wheelers overhauled and fitted with electric lighting.

An odd feature of the Barry company's coaches was that the Thirds and Composites were built with a different style of panelling to the Brake Thirds, made clear by the Gloucester Railway Carriage and Wagon Company photographs. When painted in GWR chocolate and cream in 1923/4 the Thirds and Composites were given some very odd lining out.

Mr Hosgood designed and built two steam motor cars, delivered in 1905, similar mechanically and in body shape to the later GWR steam railmotors but with fixed ventilators above the windows and at first with lattice gates on the vestibules. They carried the new dragon crests, but no class lettering. In May 1905

Class E No. 33. The Class E was initially introduced for light shunting work within the docks area. However, in 1909, No. 33 was converted to an 0-4-2T wheel arrangement and fitted for auto working on passenger services between Barry and Llantwit Major. It was coupled to the "Motor Set" comprising a four-wheel composite and a six-wheel brake third. The driver's compartment in the brake end was not, however, fitted with a regulator. The driver communicated with the fireman in the locomotive by means of wires fitted across the coach roofs. These can be seen in the picture between the coach and the locomotive. LOCOMOTIVE PUBLISHING CO. REAL PHOTOGRAPHS 25002

they were advertised as taking over the whole service between Pontypridd and Cardiff, with two new halts opened but this was over ambitious for only two vehicles, and lasted a very short time. They were later used on the Vale of Glamorgan line, well within their capacity.

Wagon stock was painted dark red and lettered "BARRY CO" in white block letters about five inches high on the second plank down on both open wagons and vans, one word to each end of the vehicle. The number was central, lower down as "NO XX". This style lasted until 1904 at least; the first batch of the Iron Mink type steel vans carried it. Goods Brake Vans were 4-wheel double-ended ten-tonners, also lettered "BARRY CO" on the third plank with the numbers on the fifth/sixth planks as "NO Q XX", the circular transfer crest being applied in the middle. This style was carried up to the last batch built in 1900. The van ends were painted bright red.

Mr Golding took over in 1905 and introduced several changes; one was to darken the locomotive livery to quite a dark red. All engines overhauled at Barry Works received new numberplates, "BARRY RAILWAY CO" in small lettering around the top edge, large numerals in the centre and date of building in small letters at the bottom. (Some photographs show "REBUILT" at the bottom edge but this was removed under Mr Auld.)

He also smartened up the livery, giving more time in the Works to painting and lining and changing the lining out from black to light green. [The difference is illustrated by the two photographs of No. 33.] He removed the brass chimney numerals and replaced them on the goods engines with small cast plates on the chimneys and on the passenger engines by gilt lettering "No XX" on the front buffer beams. This meant that goods engines carried "B R" on both buffer beams whilst passenger engines had it on the rear beam only.

Carriages were brightened by introducing large class numerals 1, 2, 3 on the doors in place of the small waist panel lettering. Brake Vans were altered, some by removal of one or both Third Class compartments, others having an extra compartment added at one end. Unfortunately, photographs are rare of all but the full van conversions.

Barry Pier station was opened on 27th July 1899 to facilitate taking passengers to the P & A Campbell steamship services to Weston-super-Mare and Ilfracombe. It had a floating pontoon (seen to the right in the picture) that connected the station to the pier. The Barry Railway did not begin its own steamer services until 1905.

FROM A CONTEMPORARY POSTCARD

Wagons also received the Golding treatment. It is likely that it was he who introduced a style of large lettering "By Co" with numbers lower down as "No" left of door, figure to right. Not all stock received this style: wagons used only at Barry Docks were left with the original "BARRY CO" lettering and carried a large "X" (see page 140).

By 1900 many railways were developing auto trains to cope with traffic created by steam Motor Cars, including the neighbouring Taff Vale and Great Western Railways. Mr Golding decided to test the idea on a modest scale. Two close-coupled sets were made up of a four-wheel Composite and a converted six-wheel Third each with electric bell control to the locomotive. Three of the smallest Barry locomotives were overhauled, repainted and given vacuum brakes and Golding style numberplates, No. 33 as auto engine was fitted with bell controls, No. 53 as spare engine without, both running as 0-4-2Ts at first. No. 50 as reserve was only given vacuum brakes but not altered. The train ran extra trips on the Vale of Glamorgan line. One of the Motor Sets had the unusual arrangement of the company title, "BARRY" on one coach and "RAILWAY" on the other, above the windows. The panels on the other set were different, it was lettered "B R" in the waist.

Mr Auld took over in January 1910 and cut back on painting and finishing to speed up overhauls. Goods engines ceased to be lined out, passenger engine lining returned to black and coupling rods between crankpin bosses were painted red. Auto train working was dropped and the Motor Set as it was still called worked normally.

In 1914 the two steam motor cars were converted to trailers and ran as a pair with corridor connection, now providing all three classes [First and Third in one and Second and Third in the other], and called the Vestibule Train in the working timetables.

Also in 1914 Mr Auld brought out his only class of locomotives, the ten Class L 0-6-4Ts. In plain red livery with the dragon crest on the bunker sides and a new pattern numberplate, lettered "BARRY RAILWAY" above the number and 1914 below together with small rectangular numberplates near the top of the smokebox doors.

A retired locomotive man at Barry, Mr J Pitt, Barry Railway cleaner and fireman from 1911 to 1922, considered the maroon livery carried by the British Railways Western Region diesel hydraulic "Western" class locomotives to be close to the later period Barry Railway colour.

Wagon livery also changed after 1910, large initials "B R" were adopted; the lettering now following GWR practice in layout. All the illustrations in the 1913 "Appendix to the Rule Book" showed the new style. Mr Auld also designed the six-wheel 20 ton Brake Van illustrated on page 112. Twelve were built in 1915, a lengthened version of the earlier four-wheelers.

Locomotive Superintendents

Date	Name	Previous employment
1888-1905	John Howell Hosgood	Taff Vale Railway
1905-1909	H F Golding	Penarth Dock, Taff Vale Railway
1909-1922	John Auld	Glasgow & South Western Railway

In addition to mineral, passenger and goods traffic, the Barry Railway also offered a service delivering newspapers and letters. Newspapers had been allowed to be transported free of charge by virtue of the Stamp Act 1712. However, in 1855, the Act was repealed and newspapers were no longer entitled to free carriage. As the Post Office did not have a parcel delivery service until 1883, the railways sensed a commercial opportunity and introduced a newspaper delivery service for which stamps were required. On the left is a Barry Railway version of just such a stamp. Different values were available and were generally charged according to the weight of the parcel.

The General Post Office had been granted a monopoly on the delivery of letters from as early as 1660. But in 1891, the Post Office signed an agreement for letters to be delivered by rail. The system in use was that the letter had to have a Post Office stamp in addition to the railway stamp. The railway stamp had to be twice the postage fee of that of the Post Office. So if the general postage was 1d, the railway letter had to be 2d. The example on the right is of a Barry Railway letter stamp. In the centre of the stamp, a number has been overprinted. This was known as the "control number" which was intended to reduce the risk of fraudulent use or theft.

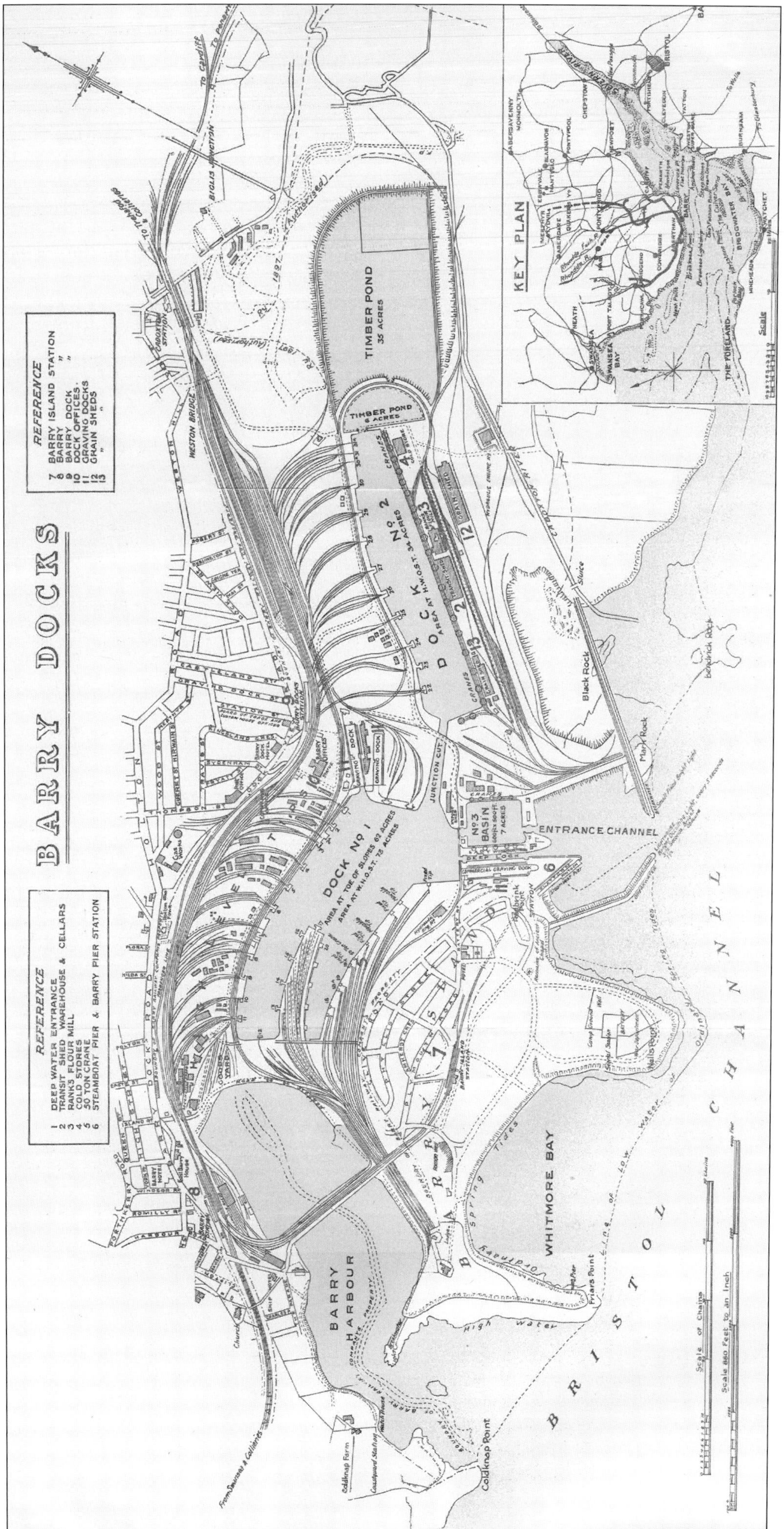
BARRY DOCKS
REFERENCE
1 DEEP WATER ENTRANCE
2 TRANSIT SHED WAREHOUSE & CELLARS
3 RANKS FLOUR MILL
4 COLD STORES
5 50 TON CRANE
6 STEAMBOAT PIER & BARRY PIER STATION
REFERENCE
7 BARRY ISLAND STATION
8 BARRY "
9 BARRY DOCK "
10 DOCK OFFICES
11 GRAVING DOCKS
12 GRAIN SHEDS
13 " "
TIMBER POND
35 ACRES
TIMBER POND
6 ACRES
DOCK No 2
AREA AT H.W.O.S.T. 34 ACRES
DOCK No 1
AREA AT TOE OF SLOPES 62 ACRES
AREA AT H.W.O.S.T. 73 ACRES
No 3 BASIN
7 ACRES
JUNCTION CUT
ENTRANCE CHANNEL
BARRY HARBOUR
WHITMORE BAY
BARRY ISLAND
BRISTOL CHANNEL
Black Rock
Bendrick Rock
Friars Point
Cold Knap Point
WESTON BRIDGE
KEY PLAN
Scale of Chains
Scale 880 Feet to an Inch

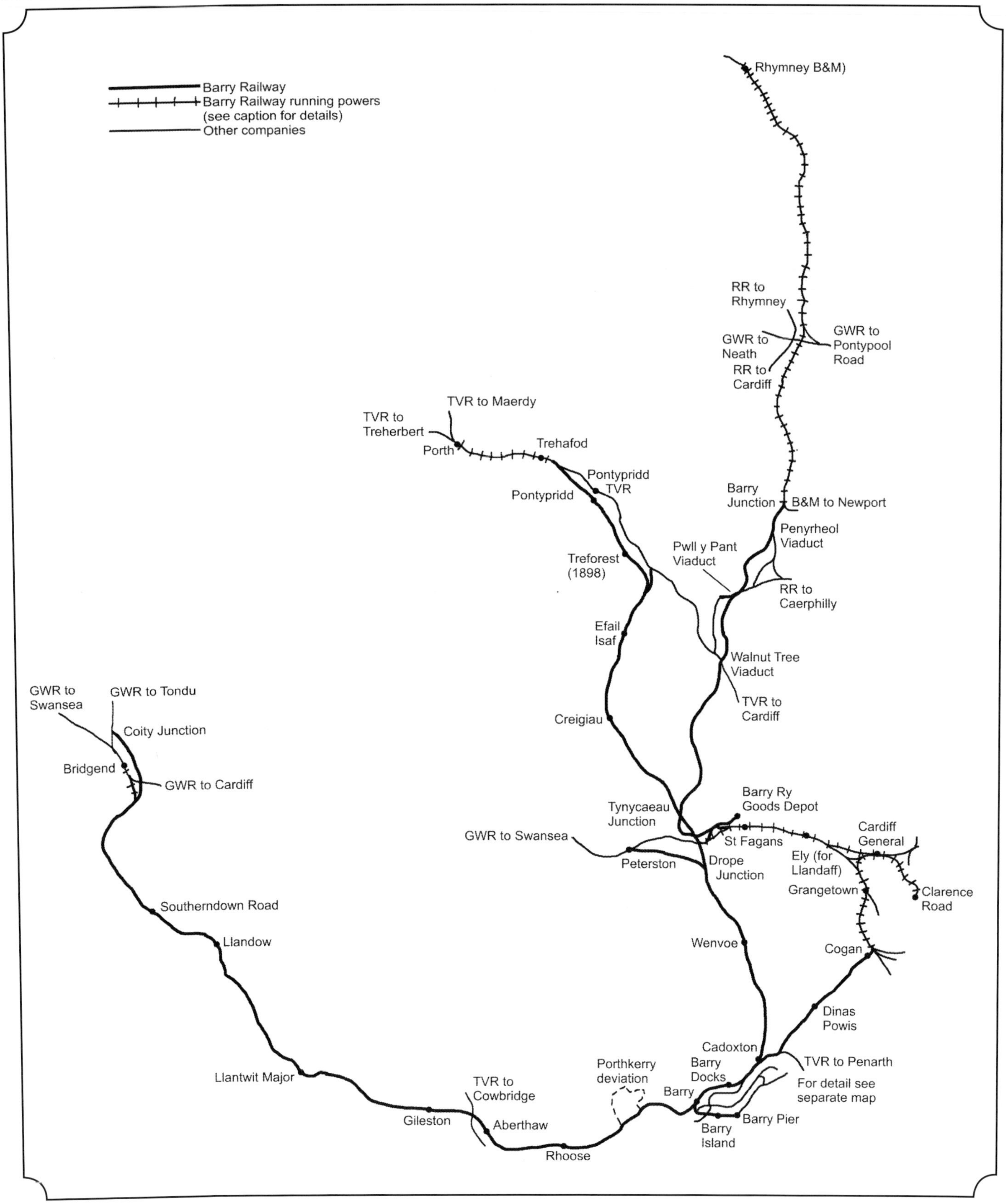

The Barry Railway system, including the Vale of Glamorgan Railway which was worked as an integral part of the Barry system
The company had running powers over other companies' lines as follows:
Brecon & Merthyr Railway: Barry Junction to Rhymney (B&M) for coal, coke and pitwood
Great Western Railway: Bridgend Junction to Bridgend station; Penarth Curve, South Junction to Cardiff, Clarence Road; St Fagans Junction to Cardiff, Clarence Road, all for coaching traffic
Taff Vale Railway: Cogan Junction to Penarth Curve, South Junction for all traffic; Trehafod Junction to Porth for coaching traffic; Penarth Curve, South Junction to Taffs Well Junction, not exercised
Not shown are two very short lived halts, Tonteg Halt and St-y-Nyll Halt, both of which opened and closed in 1905.

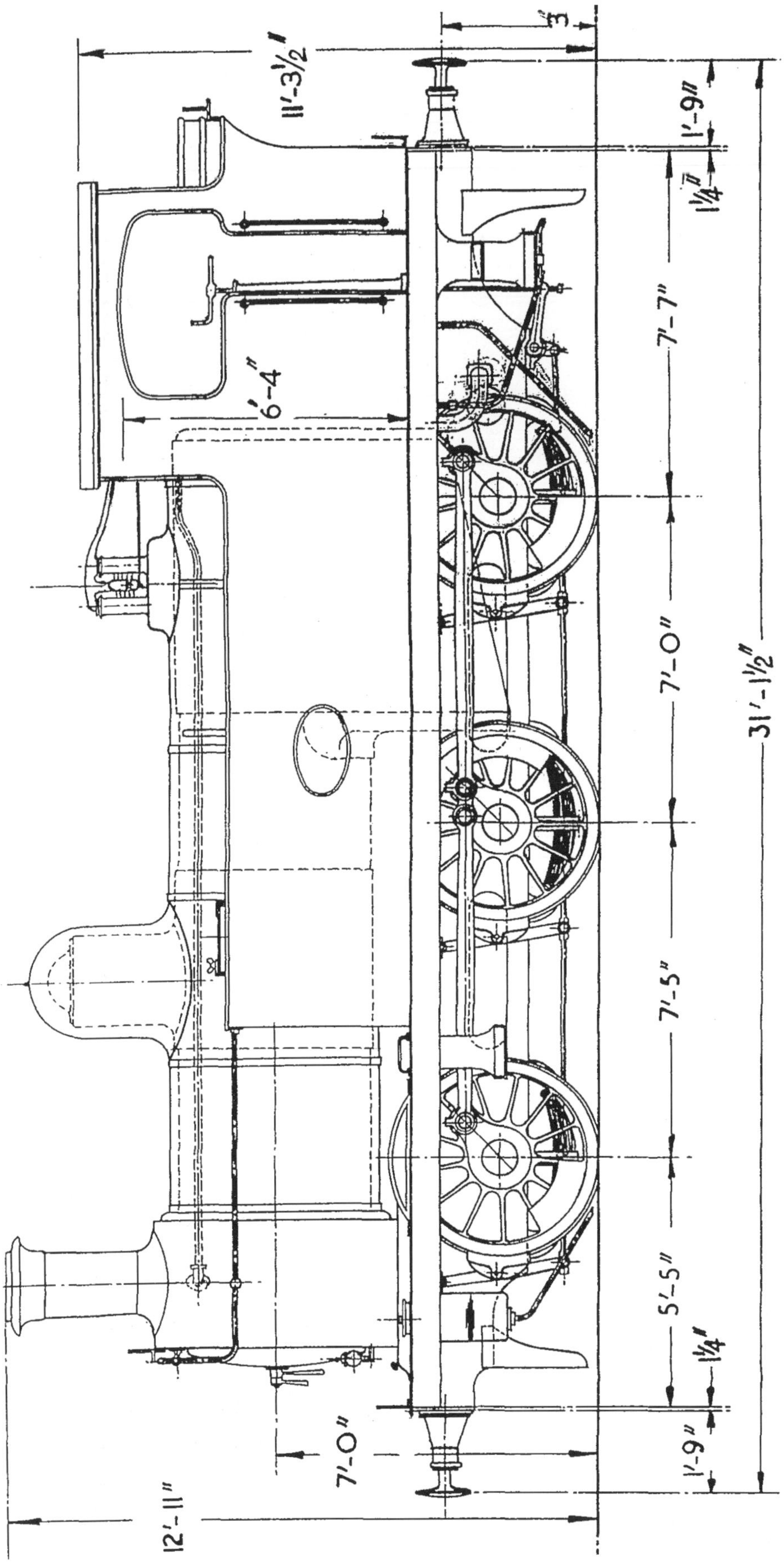

Drawing BRL1: 0-6-0T Class A Locomotive of 1888

The first locomotives to be delivered to the Barry Railway were five side-tank engines intended for shunting duties at Barry Docks. However, before the opening of the docks, a passenger service was started between Barry Docks and Cogan in 1888, a journey of five miles. Thus one of the locomotives, No. 5, was fitted with a vacuum ejector and train pipes in order to run that service. The other four had steam brakes only. With the arrival of the Class C passenger locomotive in 1889, No. 5 was relieved of its passenger duties and consequently the vacuum brake fittings were removed.

As with many of the Barry locomotives, they had inside frames and inside cylinders. Other features were a lever reverse and Allan straight link valve gear, the latter being standard practice on the Barry but not often used in the rest of the UK. Other subsequent changes included the replacement of the Type 1 two-ring boiler with the Type 2A three-ring boiler between 1898 and 1900, and the addition of coal rails probably at the same time. During Mr H F Golding's tenure as locomotive superintendent between 1905 and 1909, tyres were thickened by ½ in. to 3½ in. and Golding's successor, John Auld, reduced the size of the cylinders to 17½ in. in diameter and fitted the Type 3A boiler.

The drawings throughout this book are reproduced at a scale of 7 mm/ft, suitable for O gauge modellers, except for certain details where stated.

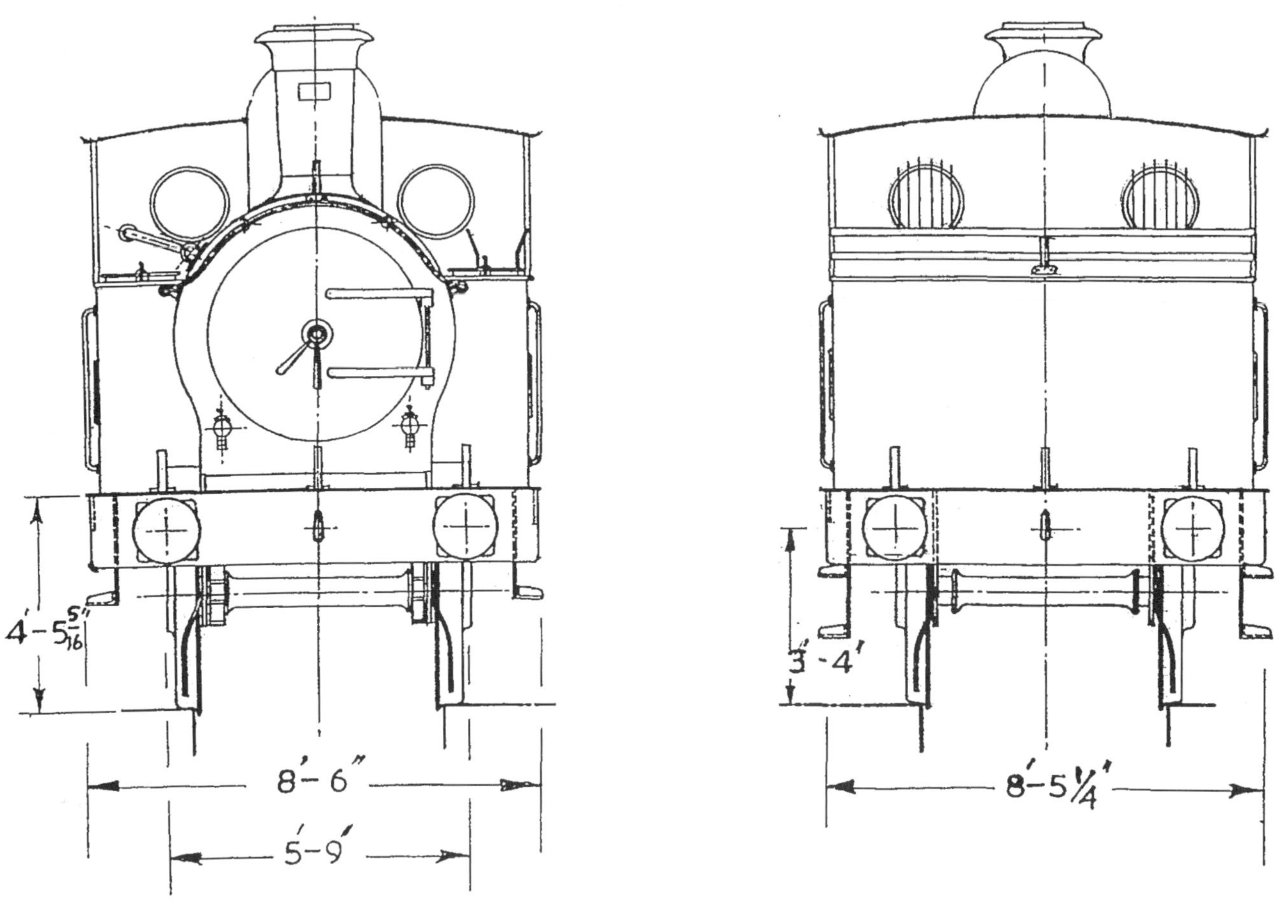

Class A No. 5. Although numbered 5, it was the first of the class to be delivered and operated the passenger service between Barry Dock and Cogan when services began in 1888, later extended to Barry in 1889. It was then allocated to heavy shunting duties in September 1889 when the Class C arrived to assume passenger duties. LOCOMOTIVE PUBLISHING CO. REAL PHOTOGRAPHS 54611

Minutes of the Board of Directors

15 June 1887
It was resolved to advertise for 15 locomotives (with the option of taking 20), to be constructed of the best materials, 6 wheels coupled, and capable of taking 500 tons nett, in all weathers, over a gradient of 1 in 400, part in tunnel, such locomotives to be constructed and delivered within 12 months, and the tenders to be sent in by the next meeting.

19 August 1887
Tenders for locomotives were opened and were referred to Mr Barry to report thereon.

17 February 1888
Messrs Sharp, Stewart & Company's plan for passenger locomotives was produced, and it was resolved to instruct the Engineer to ask for specification and price for the same.

It was resolved to order five shunting engines without pony wheels under the terms of the original contract with Sharp, Stewart & Company.

16 March 1888
Messrs Sharp, Stewart & Company's letter to Mr Wolfe Barry upon the subject of shunting engines and passenger engines was read offering to construct two passenger engines for £1,645 and five shunting engines for £1,595 each, and it was resolved to give them the order provided they accept £20 for each shunting engine off the price named, this without prejudice to the Company's position under the original contract.

Sales to industry of Barry Railway Class A locomotives

GW No.	Former BR No.	Date of sale	Sold to	Location	Scrapped
699	1	06/32	J F Wake for Coltness Iron Co.	New Mains, Scotland	01/62
700	2	10/27	Guest, Keen & Nettlefolds	Dowlais Steelworks	08/50
703	4	08/32	Ocean Coal Co. Ltd	Lady Windsor Colliery, Ynysybwl	1956

Class A No. 1. The purpose of this class was to undertake heavy shunting duties at the docks. In this particular photograph can be seen the old style number plate with the word "Barry" above the number. However the new device of a Welsh dragon under a stag's head can be seen on the bunker, introduced by H F Golding after 1905. Ken Nunn collection H1011/LCGB on 26 April 1910

Class A Specification	
Maker	Sharp, Stewart
Numbers	1–5
Maker's numbers	3449–3453
Date built	11/1888
Wheel arrangement	0-6-0T
Cylinders	18 in. × 26 in.
Wheels	4 ft 3 in.
Wheelbase	7 ft 5 in. + 7 ft 0 in.
Tractive effort	21,060 lb
Bunker capacity	30 cwt
Tank capacity	1,000 gallons
Weight full	45 tons 6 cwt

No. 1 again, this time shown with the enginemen and the shunter complete with pole to the rear of the locomotive. Being considered a freight engine, it was not lined out as were the passenger engines. D Davies collection

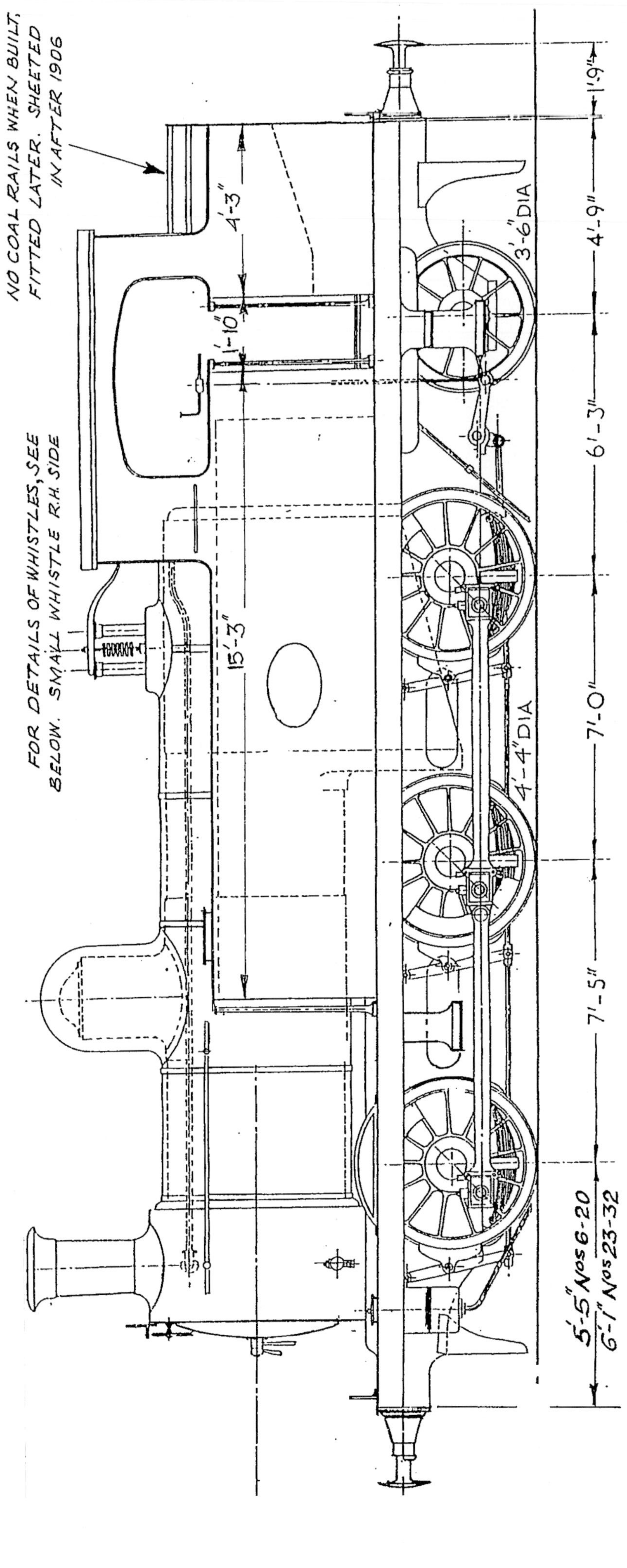

DRAWING BRL2: 0-6-2T CLASS B LOCOMOTIVE

Introduced for the haulage of main line coal trains, the Class B was virtually a carbon copy of the Class A with the addition of a trailing axle used to support a larger bunker needed for the longer journey from Barry to Trehafod. Three of these locomotives arrived in 1888 followed by another 12 in 1889 before the opening of the main line in July of that year. Subsequent to the opening, another 10 were delivered, the last being in February 1890, resulting in a class of 25.

The bunker capacity was 40 cwt and the tank capacity was 1,400 gallons. This was to prove significant in traffic, as the class was unable to reach Trehafod without the need to refill the water tanks.

An early distinction was made between Nos. 6–20 which were fitted with Type 1 boilers and Nos. 23–32 which were fitted with Type 2A boilers, the former being called Class B and the latter B1. The only distinction between these boilers was that a Type 1 had two rings and a Type 2A had three. By 1903, however, all locomotives of this class were fitted with Type 2A boilers and the distinction was dropped, reverting to Class B.

Another feature of the Class B was that they were fitted with steam brakes only, with the exception of No. 7, which was assigned the role of relief passenger locomotive for No. 5. This equipment was removed after the arrival of the Class C and it returned to main line freight duties.

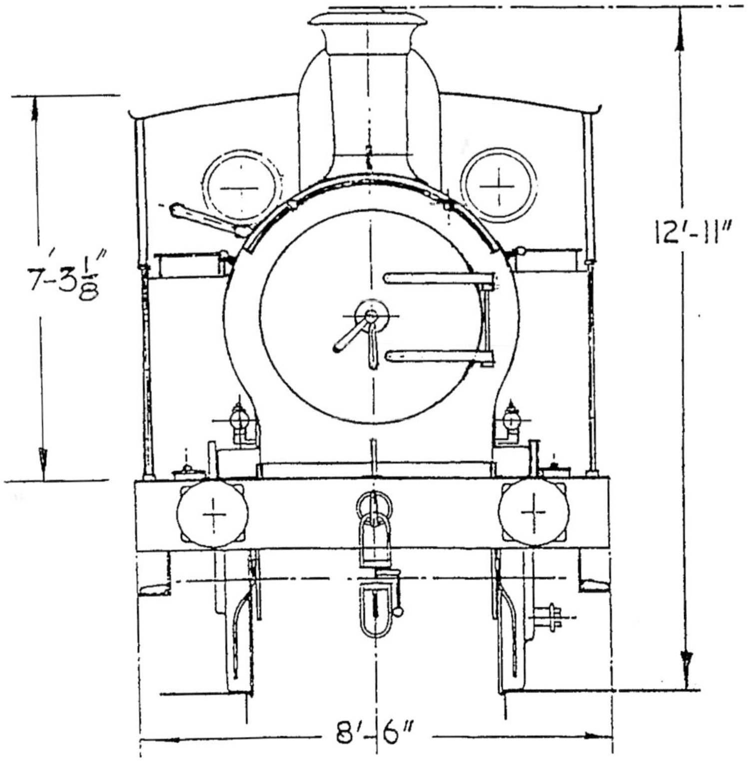

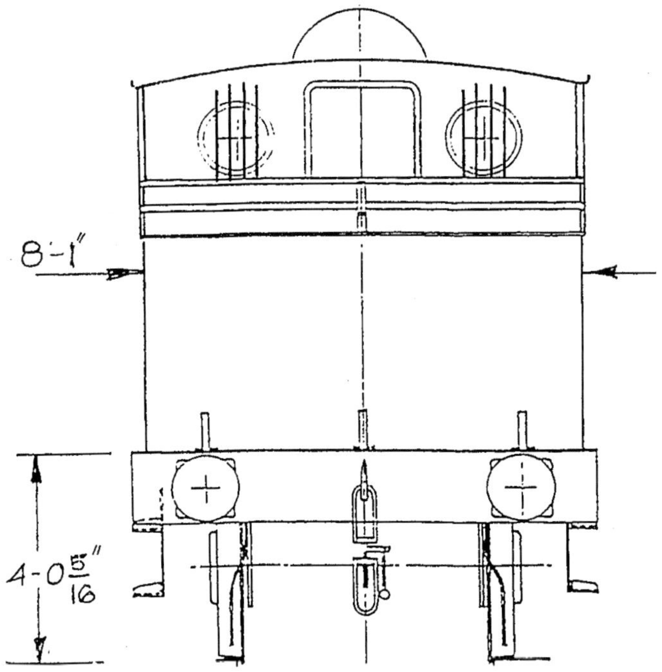

Sales to industry of Barry Railway Class B locomotives

GW No.	Former BR No.	Date of sale	Sold to	Location	Scrapped
227	27	10/34	Robert Frazer & Sons Ltd	Hartley Main Collieries	09/60

Class B No. 25. This class was the first of the freight locomotives to be used for all mineral and goods traffic hauled outside of the docks area. It was delivered in January 1890 and, because it had a Type 2A boiler, was given the designation "B1". This was dropped when the subsequent Class B1 arrived later in the same year. C W Harris collection

Class B No. 32. This was the last of the class to arrive in February 1890 and carried the same B1 designation as No. 25. After the arrival of the Class B1, they were reassigned to shorter distance work away from the main line. Bronson Mills, C W Harris collection

Class B No. 13. This was among the first of the class to arrive, this particular locomotive being delivered in March 1889. It was fitted with a Type 1 boiler and therefore carried the categorisation "B" in contrast to subsequent locomotives. One of the early batch, No. 7, was vacuum fitted and was the relief passenger engine for Class A No. 5. SHARP, STEWART & CO

CLASS B SPECIFICATION

Maker	Sharp, Stewart	Sharp, Stewart
Numbers	6–20	23–32
Maker's numbers	3454–3468	3571–3580
Date built	1888/89	1889/90
Wheel arrangement	0-6-2T	
Cylinders	18 in. × 26 in.	
Wheels		
Coupled	4 ft 3 in.	
Trailing	3 ft 6 in.	
Wheelbase	7 ft 5 in. + 7 ft 0 in. + 6 ft 3 in.	
Tractive effort	21,060 lb	
Bunker capacity	40 cwt	
Tank capacity	1400 gallons	
Weight full	51 tons 8 cwt	

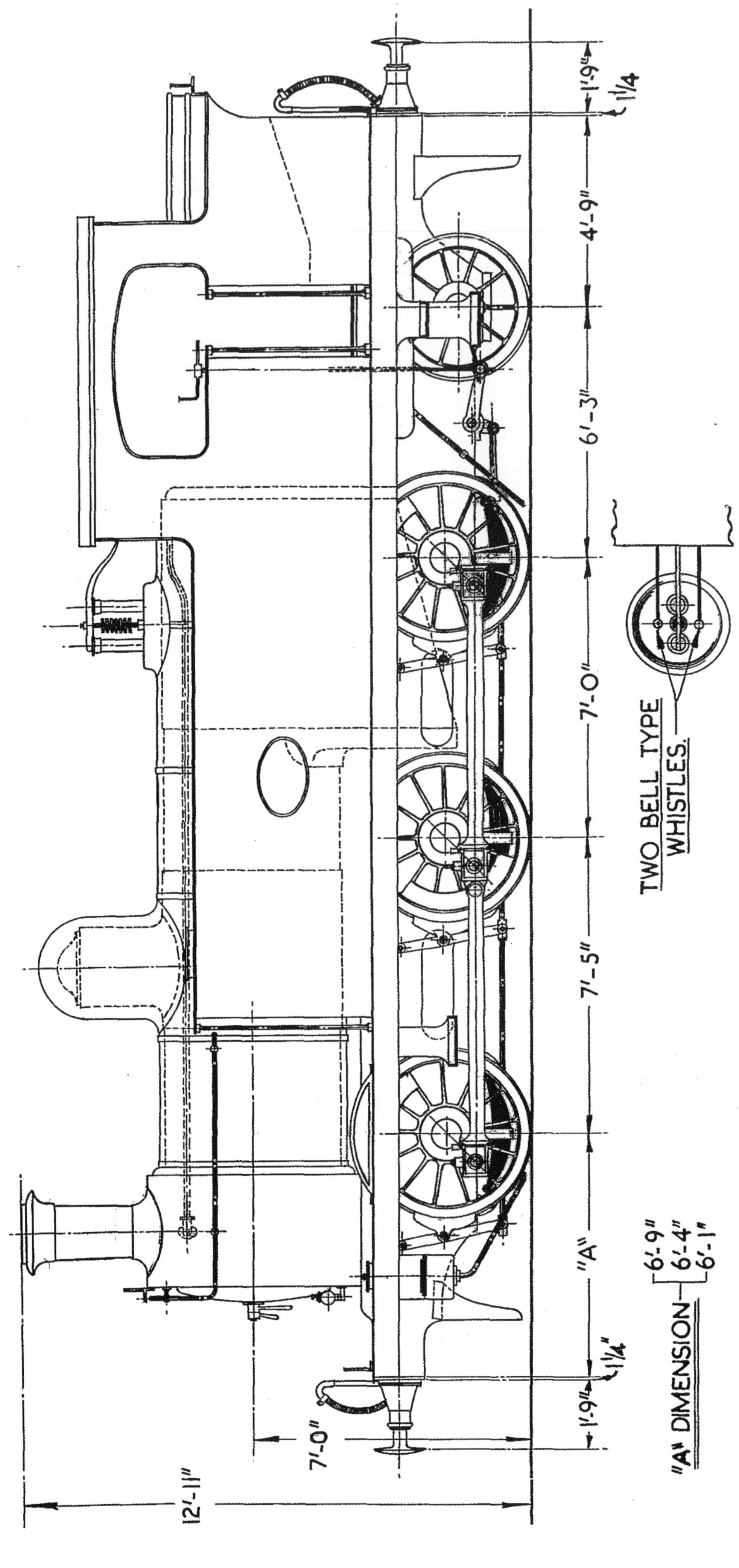

Board of Directors
17 April 1891

Tenders for the supply of locomotives and a summary thereof were submitted and considered. Mr Robinson of Messrs Sharp, Stewart & Company and Mr Gooch for the Vulcan Foundry Company respectively appeared before the Board, and stated what deductions in the prices they were prepared to make in the event of the number of locomotives ordered being increased. After a careful consideration of the figures submitted by these gentlemen, it was resolved that the tenders of the Vulcan Foundry Co Ltd be accepted, as follows, viz:-

For two passenger engines at £2,060 each;
For two saddle tank engines at £1,705 each;
For ten side tank engines at £1,940 each.

Mr Gooch, before leaving, wrote out and deposited amended tenders embodying the above prices, and undertook that delivery of the first engine should be made by 15th December 1891, the rest to follow at a rate of one per week.

Drawing BRL3: 0-6-2T Class B1 Locomotive

With 42 locomotives in this class, this was the main workhorse for freight and mineral traffic. It was a variation of the Class B though its original designation was that of B2 (see Class B). The inability of the Class B to reach Trehafod from Barry without the need to replenish the water tanks prompted the production of the Class B1.

The essential difference between this locomotive and its predecessor was the water capacity and, to a lesser extent, the bunker capacity. On the B1, the water capacity was increased by 16% to 1,630 gallons and the bunker by 8% to 43 cwt.

For a short time, certain modifications were made to this class. In 1902, Hosgood replaced the slide valves on No. 59 with piston valves to improve the locomotive's efficiency. The expected benefits did not materialise and so slide valves were reinstated.

However, the B1 was not confined to freight and mineral traffic. Excursions to Barry Island as well as workmen's trains would usually be hauled by this class. In particular, two locomotives, Nos. 60 and 108, were fitted with Westinghouse braking in 1908 for working the "Ports to Ports" express between Barry and Cardiff, with the same modifications being made to Nos. 111 and 122 in April 1920 in time for the National Eisteddfod held in Barry in the same year. The Westinghouse apparatus was not judged a success and was soon removed, the coaching stock used for the services being, in any case, dual-braked.

The "Ports to Ports" express, the Barry company's only involvement in long distance services, ran from Barry to Newcastle-upon-Tyne, principally for seamen transferring to and from ships at the East Coast ports. However, the fitting out of Class B1 locomotives for Ports to Ports duties made them suitable replacements on passenger services for the Class G and Class J when they were being repaired.

Later in 1910, two Class B1 locomotives, Nos. 38 and 43, were fitted with steam-heating apparatus and were vacuum-fitted for the same Ports to Ports service, which they worked to and from Cardiff. In contrast, steam heating did not exist on the Barry Railway's own trains. The company's view was that, journeys on its network being short, passenger heating was not worth the bother.

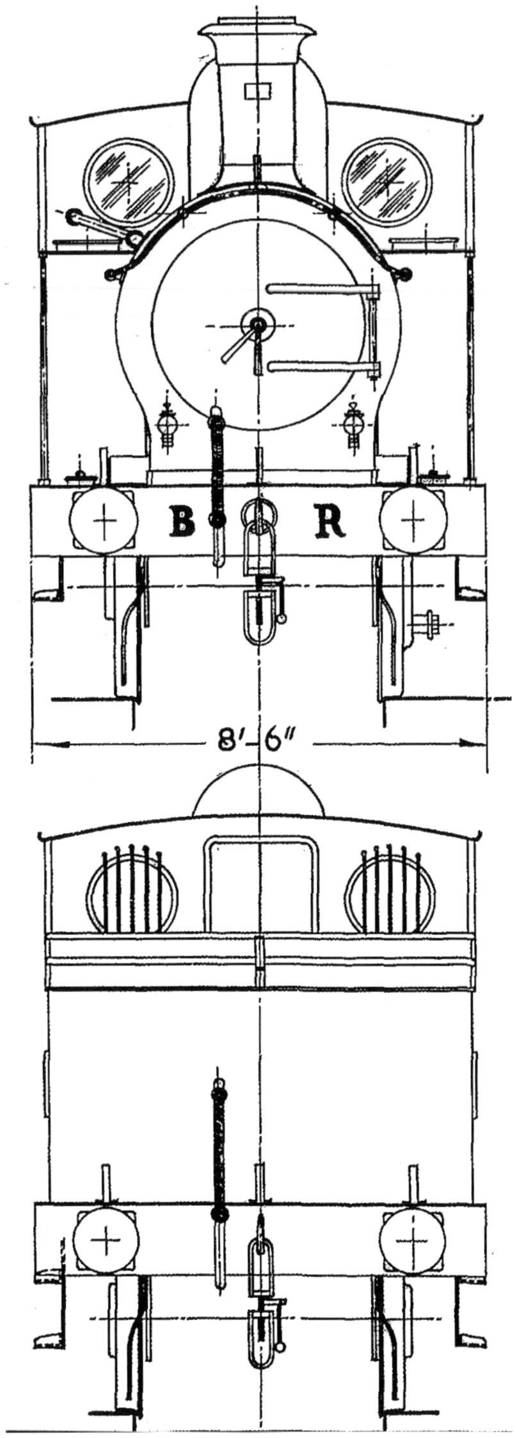

Class B1 No. 109. Altogether, three companies were involved in the construction of this class, Sharp, Stewart, Vulcan Foundry and S A Franco-Belge. The locomotive in the photograph is from the third batch constructed by Sharp, Stewart and delivered in June 1900. It survived into the nationalisation era and was withdrawn by British Railways in November 1949.
C W Harris collection

No. 112 was part of the same batch as No. 109. It is shown at the head of a goods train, of which there are very few photographs of the Barry Railway. It also survived into the nationalisation era and was withdrawn in April 1948.
C W Harris collection

Class B1 No. 76. This locomotive arrived as part of the second batch from Sharp, Stewart in December 1894. Its type 2A boiler was exchanged for a second-hand type 3A boiler in October 1921. This locomotive also survived into the nationalisation era and was withdrawn in November 1949.
Locomotive & General Railway Photographs ref 23233

Class B1 Specification

Maker	Sharp, Stewart	Vulcan Foundry	Sharp, Stewart	Sharp, Stewart	S A Franco-Belge
Numbers	38–46	54–63	73–78	105–116	122–126
Maker's numbers	3598–3606	1336–1345	4044–4049	4607–4618	1272–1276
Date built	1890	1892	1894	1900	1900
Wheel arrangement	0-6-2T				
Cylinders	18 in. × 26 in.				
Wheels					
Driving	4 ft 3 in.				
Trailing	3 ft 6 in.				
Wheelbase	7 ft 5 in. + 7 ft 0 in. + 6 ft 3 in.				
Tractive effort	21,060 lb				
Bunker capacity	40 cwt	32 cwt	43 cwt	43 cwt	43 cwt
Tank capacity	1630 gallons	1750 gallons	1630 gallons	1630 gallons	1630 gallons
Weight full	55 tons 18 cwt	56 tons 10 cwt	55 tons 18 cwt	57 tons	57 tons

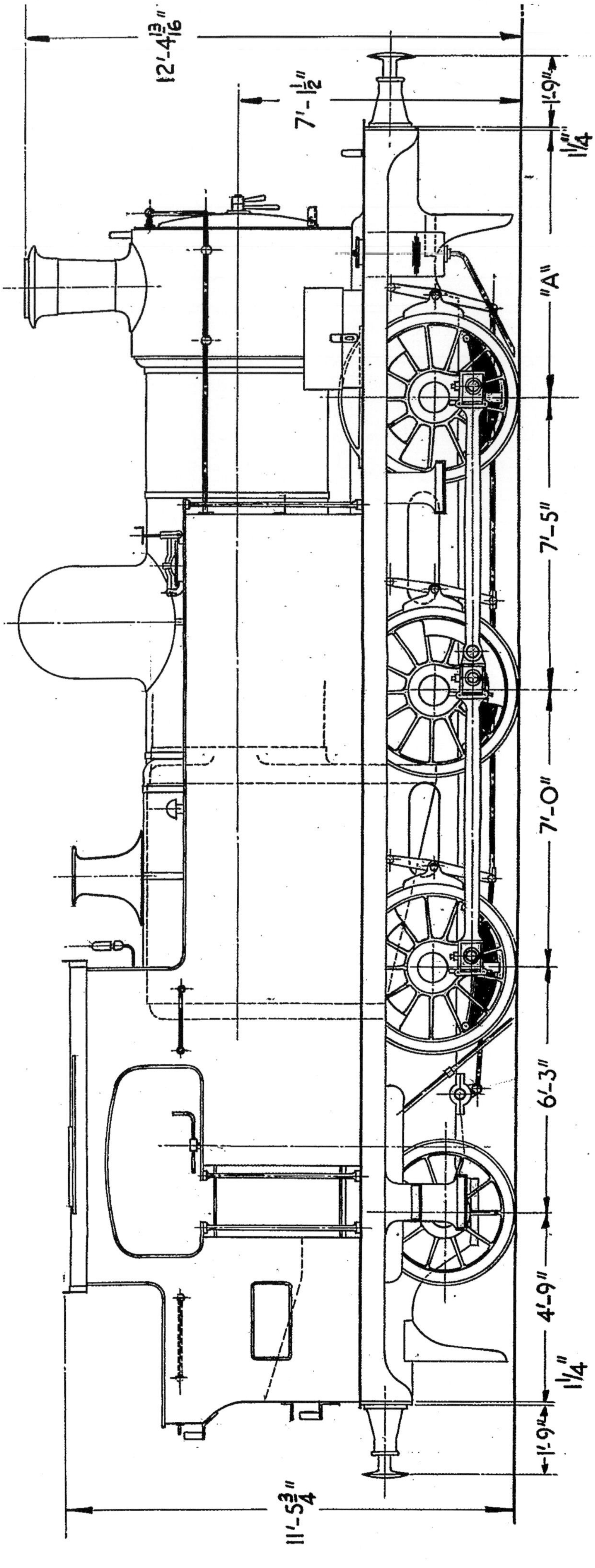

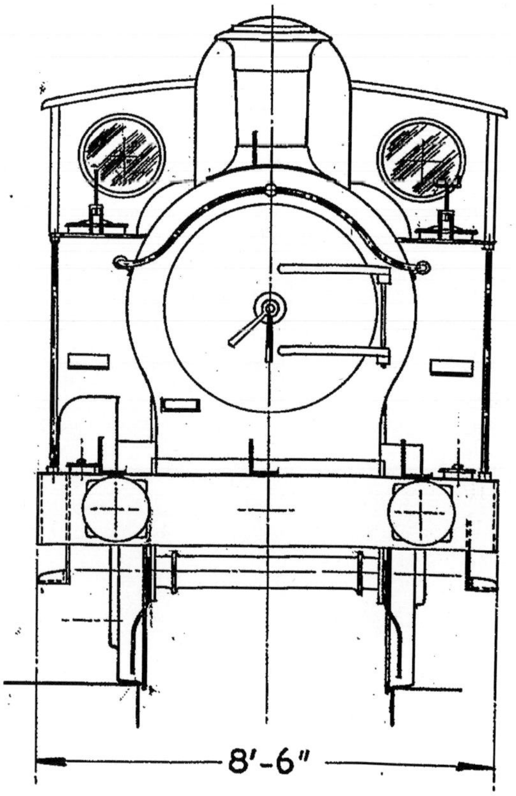

Note
Dimension A was 6 ft 9 in., 6 ft 4 in., and 6 ft 1 in. on different batches.

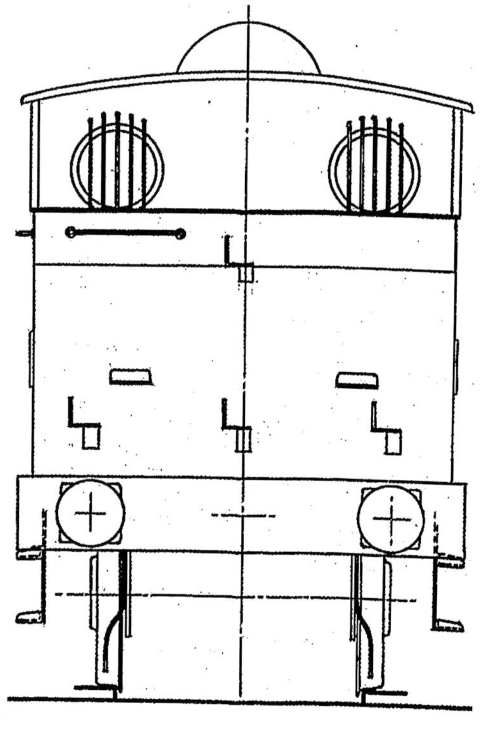

Drawing BRL4: 0-6-2T Class B1 Locomotive as rebuilt by the GWR

After the Barry locomotives passed to the GWR in 1923, some were sent to Swindon in 1924 for renumbering and modifications. In the case of the B1, this involved a more even distribution (14t 15c + 14t 15c + 14t 9c + 9t 10c) of the engines' 53 ton 9 cwt total weight. In 1924, on five of the engines (Nos. 233, 240, 249, 250 and 275), standard No. 9 superheated domed boilers were fitted and tractive effort increased to 21,230 lb. No. 267 was modified in the same manner in 1927 followed, starting in 1928, by another 16 of the class being modified at Caerphilly.

There were numerous boiler changes during the GWR era though three (269, 274 and 276) were fitted with non-superheated standard No. 9 boilers in January 1932, January 1931 and August 1940, respectively.

Two of the locomotives (250 and 268), were fitted with high-roofed cabs, GW bunkers which increased the coal capacity to 55 cwt and GW side tanks which reduced the water capacity to 1,600 gallons.

Initially, the class was allocated to the duties it had carried out on the Barry. But from 1925 onwards, the GW 56xx class began to take over and replace the B1 on the main line. From 1932, the vacuum brake fittings were removed and the locomotives were allocated to shunting at Barry and Radyr.

Out of a class of 42 locomotives, 18 survived into the nationalisation era and were given the British Railways power classification of 3F.

Renumbering

Barry Railway numbers	Great Western numbers
38–40	233–235
41	238
42–46	240–244
54–63	245–254
73–78	255–260
105–116	261–272
122–126	273–277

Sales to industry of Barry Railway Class B1 locomotives

GW No.	Former BR No.	Date of sale	Sold to	Location	Scrapped
244	46	07/39	J F Wake for Manchester Collieries Ltd	Walkden Yard	06/52
251	60	07/34	Lilleshall Iron Co	Oakengates	01/57

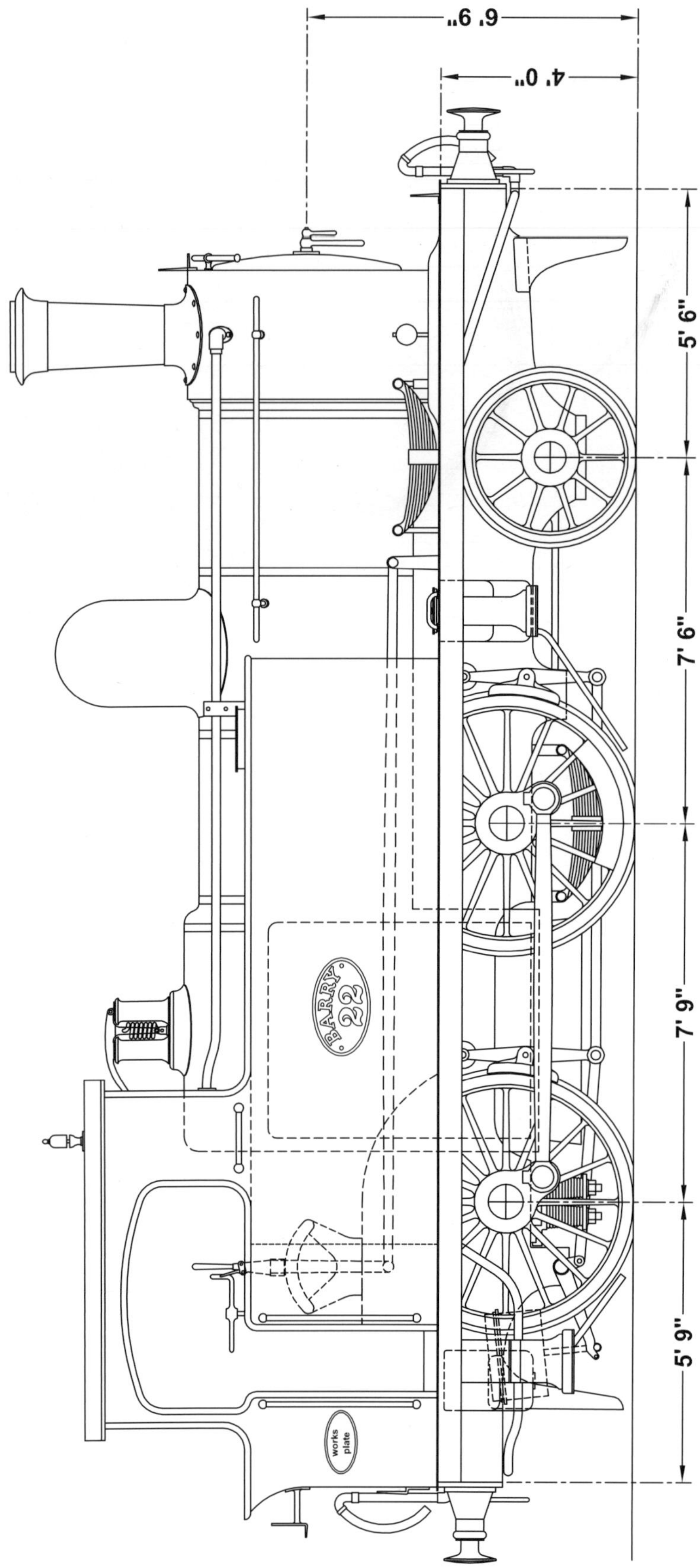

Drawing BRL5: 2-4-0T Class C Locomotive

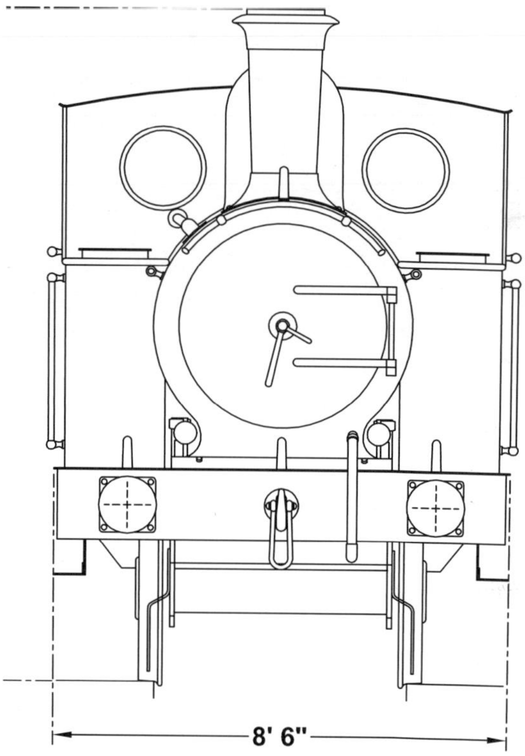

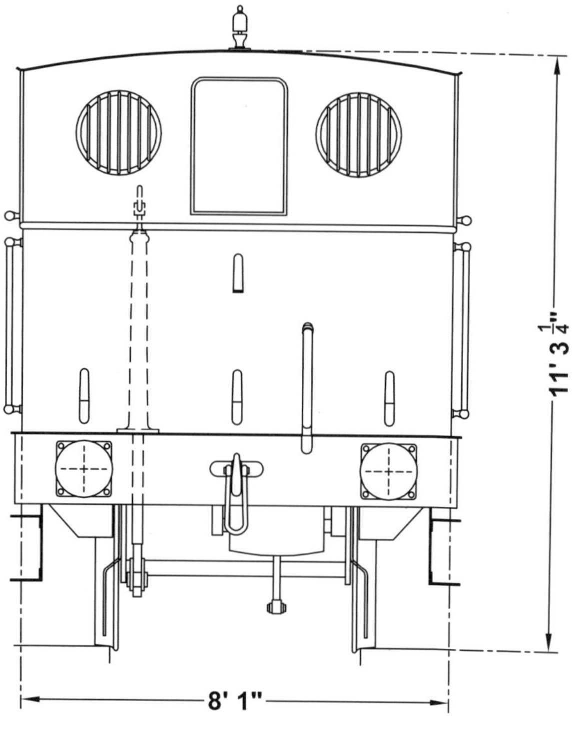

Drawings by Richard Evans

Class C No. 22. Shown in its amended 2-4-2T wheel arrangement, No. 22 is at Cardiff Riverside station facing south with an ex-Lancashire & Yorkshire Brake Third behind. Trains either terminated at Cardiff Riverside or proceeded to Cardiff Clarence Road. CLIFFORD COLLECTION NO. 26

Class C Specification	
Maker	Sharp, Stewart
Numbers	(a) 21, 22; (b) 37, 52
Maker's numbers	3528, 3529, 3610. 3626
Date built	(a) 9/1889; (b) 9/1890
Wheel arrangement	2-4-0T (1898 Nos. 21, 22 and 52 2-4-2T)
Cylinders	17 in. × 24 in.
Wheels	
Coupled	5 ft 3 in.
Leading and radial	3 ft 6 in.
Wheelbase	2-4-0T 7 ft 6 in. + 7 ft 9 in. 2-4-2T 7 ft 6 in. + 7 ft 9 in. + 6 ft 6 in.
Tractive effort	14,040 lb
Bunker capacity	2-4-0T not known; 2-4-2T 45 cwt
Tank capacity	2-4-0T 800 gallons; 2-4-2T 1370 gallons
Weight	2-4-0T 41 tons 2 cwt; 2-4-2T 50 tons 15 cwt

Drawing BRL6: 2-4-0T Class C Locomotive As rebuilt as a 2-4-2T

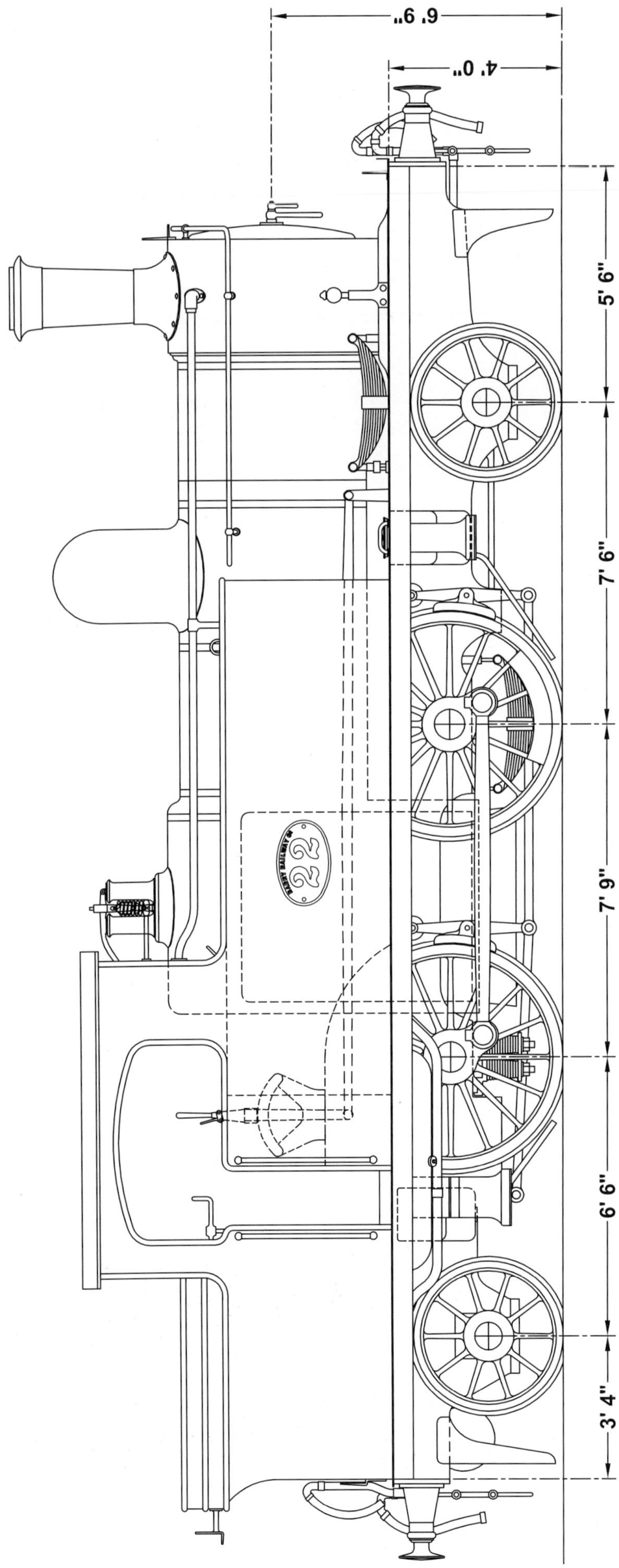

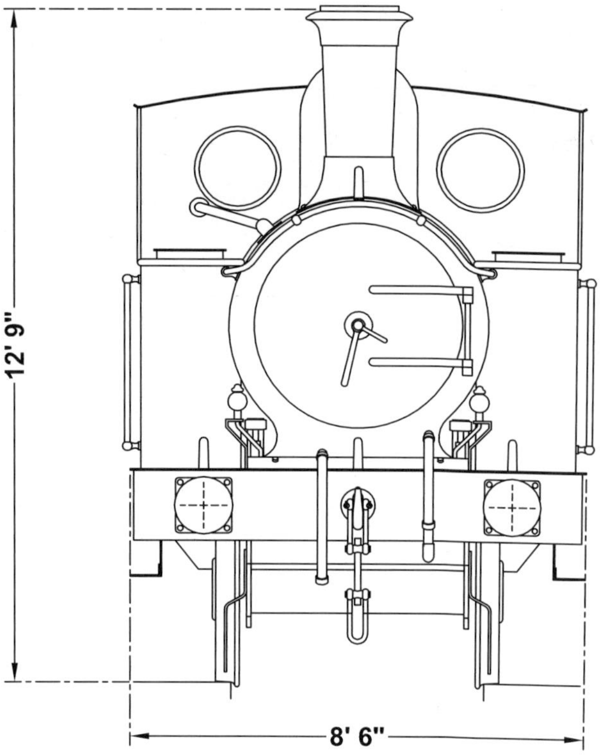

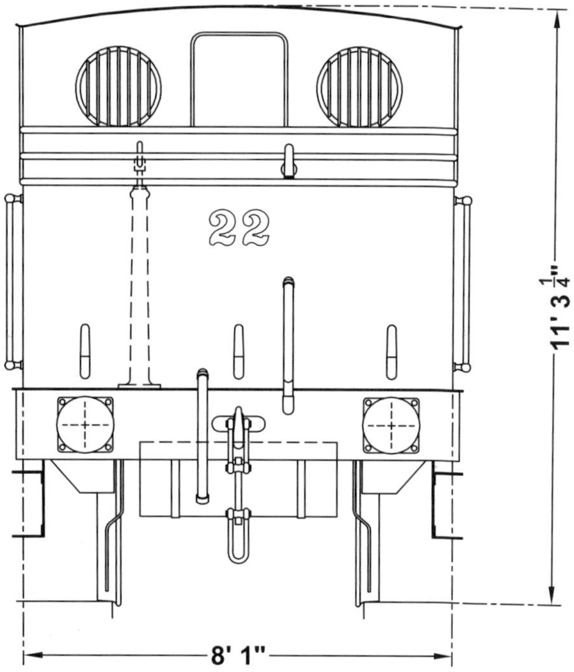

Drawings by Richard Evans

The Class C was the first Barry class specifically designed for passenger traffic and in the years in which they were delivered, 1889–1890, the Barry passenger service was confined to running between Barry Town and Cogan. There was no main line service from Barry to Porth or Cardiff service from Pontypridd and it would be another eight years before the opening of the Vale of Glamorgan Railway. Therefore the demands placed upon Barry passenger locomotives could not be described as onerous.

When the Barry Railway joined the Taff Vale Railway at Cogan, the junction allowed for the Barry–Cogan service to be extended to Cardiff Riverside and subsequently Clarence Road, and the Class C handled this traffic until 1897. This service began on 3rd August 1893 although freight traffic had made use of this connection prior to this date.

In 1898, two of the class, Nos. 21 and 22, were altered from a 2-4-0T to a 2-4-2T arrangement although the Class C designation remained. This enabled the enlargement of both the coal bunker to 45 cwt and the water storage capacity of the locomotive from 800 gallons to 1,370 gallons.

On return to traffic duties in 1898, the altered locomotives briefly returned to the Barry–Clarence Road service but were soon displaced by the third delivery of the Class J in 1899. As a result, they were relegated from principal services to minor duties. No. 21, for example was allocated to hauling both the Directors' saloon and that of the Engineer, as well as the Manager's truck. By way of contrast, No. 22 hauled colliers' trains and some Vale of Glamorgan passenger services.

Some years later, in 1914, the two unsuccessful Motor Cars were converted into semi-corridor conventional coaches and ran together as the "Vestibule Set" on the Vale of Glamorgan line between Barry and Bridgend. Nos. 21 and 22 were the locos allocated to haul this formation, whereas Nos. 37 and 52 were sold to the Port Talbot Railway in 1898. Interestingly, these numbers were reallocated to the Class F when the fourth delivery of this class arrived in 1900.

Board of Directors
25 July 1898
The secretary reported that the Port Talbot Company wished to know the price that would be asked for locomotive No. 37 at present hired to them.

Mr Hosgood's letter giving the present value as from £1,000 to £1,100 was taken as read.

It was resolved that £1,100 be asked.

Cadoxton station fire. In the early morning of Monday 27th March 1911 a fire, which originated in the booking office, completely gutted the station building on the up platform at Cadoxton station. The photograph also reveals the construction method of overlapping planked wood which was a characteristic of all the stations from Barry to Cogan. PHOTOGRAPHER UNKNOWN

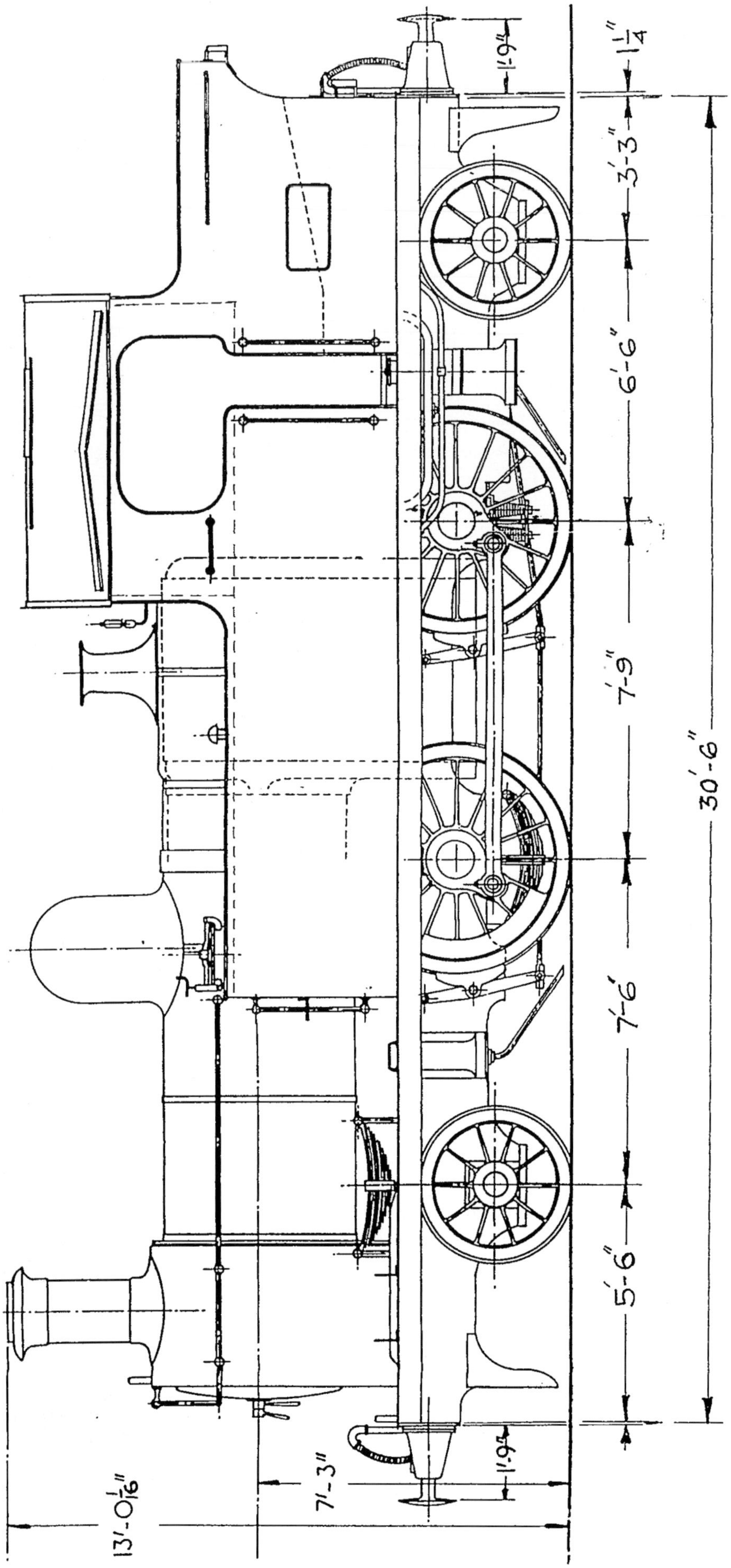

Drawing BRL7: 2-4-2T Class C as rebuilt by the GWR

Only two of the class were taken into Great Western ownership directly from the Barry: Nos. 21 and 22. Nos. 37 and 52 became GW locomotives by way of the Port Talbot Railway which had purchased the engines in 1898 for passenger services. Nos. 21 and 22 had been altered by the Barry from a 2-4-0T arrangement to that of a 2-4-2T in the summer of 1898. No. 37 (also PTR No. 37) was sold to the Port Talbot as a 2-4-0T and No. 52 (PTR No. 36) was offered by the Barry with the same wheel arrangement. However the Port Talbot decided to pay an additional £350 for the supply and fitting of an additional set of trailing wheels just as had been done on the Barry engines. The work on PTR No. 36 was carried out in November 1898.

Soon after purchase, it was clear that the engines were under-performing and so Nos. 37 and 36 were reboilered in 1901, resulting in a higher pressure and gave the locomotives a tractive effort of 14,975 lb, an increase of approximately 6.5%.

The modifications carried out by the GWR to No. 21 were not insubstantial. This locomotive (GWR No. 1322) was sent to Swindon in 1923 where it received a "Metro" boiler, a high-roofed cab and enlarged coal bunker. By way of contrast, No. 22 (GWR No. 1323) received no modifications and was sent to Swindon in 1925 where it remained until condemned in August 1926.

Although the Port Talbot Railway was not officially absorbed until 1922, the GWR took over the running of the PTR under a working arrangement and running powers agreement in 1908, which included the use of PTR locomotives. PTR No. 37 (GWR No. 1189) received a "Metro" Belpaire boiler in 1915 but it was not until Great Western days that PTR No. 36 (GWR No. 1326) was fitted with the same type of boiler in 1925.

Prior to this, however, both locomotives found themselves being replaced by GW 2-6-2Ts from as early as 1908. After its reboilering, PTR No. 37 was used as a station pilot at Swindon whereas PTR No. 36 was allocated to passenger duties between Llanelli and Pontarddulais and further to Llandovery.

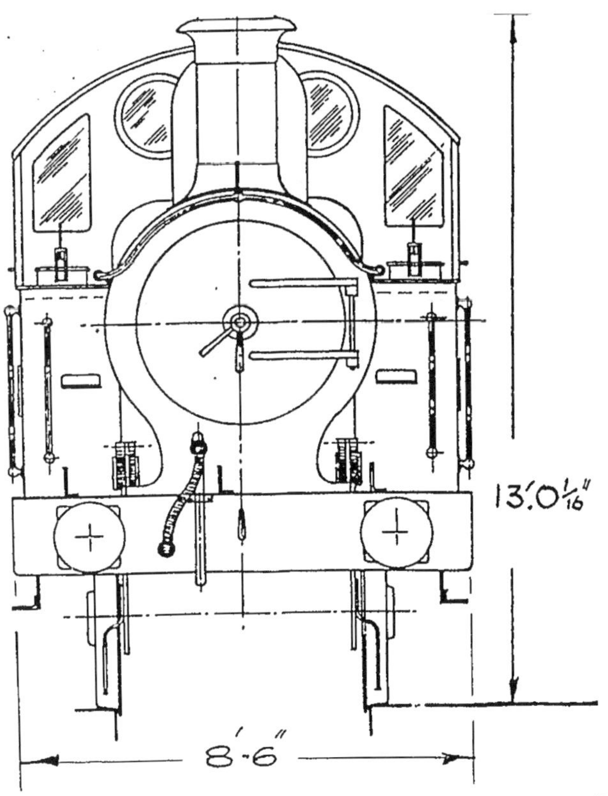

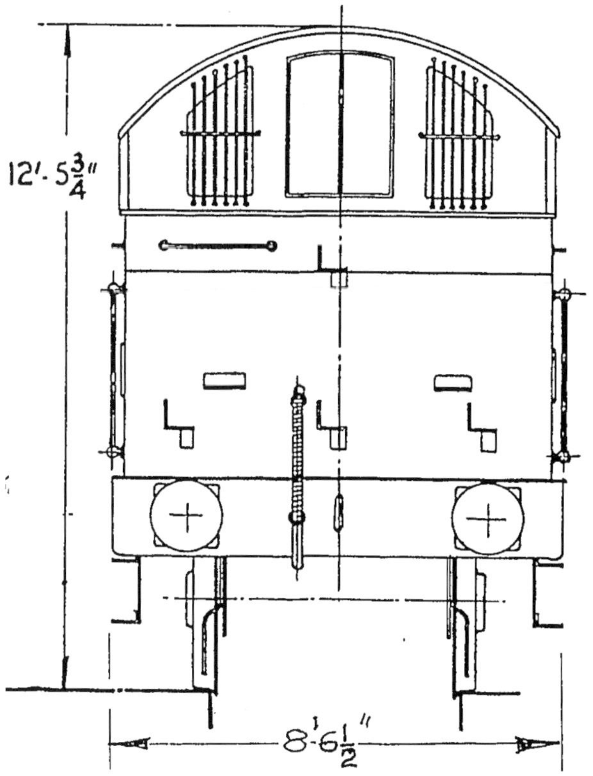

DETAIL OF NUMBER PLATE
SCALE 14 MM TO 1 FT

Class J No. 86 pulls into Riverside station with a five carriage rake of six-wheeled coaches. The locomotive at the head of the train was built by Hudswell, Clarke and delivered in 1897. Like the Class G before them, they sported brass chimney caps, dome covers and safety valve bases all smartly polished.
KEN NUNN COLLECTION 1548/LCGB

Class C No. 22 stands in the Down platform at Barry on light passenger duty with a train for the Vale of Glamorgan line. This locomotive was originally intended for auto-working with the Motor Set but Golding, the Locomotive Superintendent, changed his mind and decided to use the Class E instead. However when the Motor Cars were converted into trailers in 1914, No. 22 was the engine of choice to haul the "Vestibule Set'" on the Barry to Bridgend service. In the photograph, Barry signal box can be seen in the background. C W HARRIS COLLECTION

Renumbering

Barry Railway numbers	Port Talbot Railway numbers	Great Western Railway numbers
21	-	1322
22	-	1323
37	37	1189
52	36	1326

Hand turnout lever. With over a 100 miles of track in the Barry Docks area, hand turnout levers were a common sight. This particular example is from Barry Docks and was photographed by Mike Morton Lloyd.

Pontypridd station. This was the most northerly of the Barry Railway stations and was completed in 1896. Passenger services from Porth on the Taff Vale Railway to Barry began on 16th March of the same year, and on 7th June 1897 a new service from Pontypridd to Cardiff via St Fagans began.

FROM A CONTEMPORARY POSTCARD

An unidentified tugboat waits to enter Barry Dock Basin (also known as Dock No. 3) from the docks side while the roller bridge which spans the entrance is retracted to admit passage. These bridges were used to carry dock lines over water where a lifting bridge or a swing bridge were not feasible. The mechanism for operating these bridges was that they were first jacked up and then pulled back into a protected area on the harbour wall. The signal box for this area of the dock can be seen behind the left end of the bridge. In the top left corner of the picture is Barry Dock Breakwater Lighthouse which was built in 1890, a year after the opening of the docks. The steam crane seen to the right of the lighthouse is being used for maintenance of the masonry section of the breakwater.

FROM A CONTEMPORARY POSTCARD

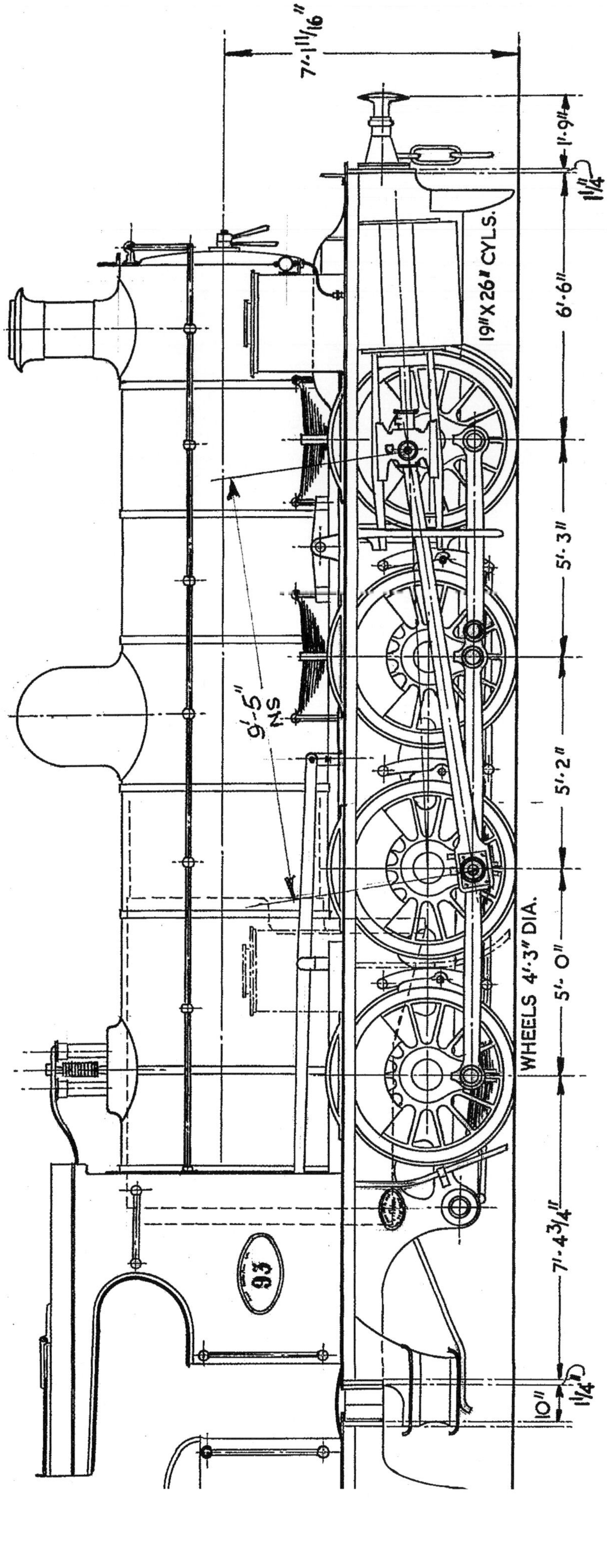

Drawing BRL8: 0-8-0 Class D Locomotive

The Class D was the only tender class ever owned by the Barry Railway and was built to a standard Sharp, Stewart design. They were originally constructed for the Swedish & Norwegian Railway which had placed an order for 20 locomotives, but that company ran into financial difficulties. As a result, two of the undelivered locos were sold by Sharp, Stewart to the Barry in 1889 becoming Nos. 35 and 36. Features of the locomotives were a Westinghouse single-acting brake, American style headlamps and cattle guards; both the headlamps and the cattle guards were subsequently removed.

Nos. 92 and 93 followed much later. Similar to the first acquisition, financial difficulties led to these locomotives being seized by bailiffs from the Swedish & Norwegian in 1891 but, due to prolonged legal action in the Swedish courts, the locomotives were not delivered to the Barry Railway until 1897. Such was the size of these locomotives that their outside cylinders had to be removed for the journey from Sharp, Stewart's works in Manchester; though they had no difficulty operating on the Barry itself, when confined to hauling coal trains between Cadoxton Yard and the docks.

One aspect of the locomotives' design that distinguished the first delivery, 35 and 36, from the second delivery, 92 and 93, was the cab sides. Nos. 35 and 36 had closed sides with a porthole window on each side. Nos. 92 and 93, examples of the earlier design of the locomotive, had open sides, resembling a U-shape placed on its side with the bottom of the U facing the front of the engine.

In 1905, the Brecon & Merthyr Extension was opened for coal traffic from the Rhymney Valley and four years later, Nos. 92 and 93 were transferred to hauling coal trains on that line. Now that their duties took them out onto the main line, their dimensions had to be adjusted to correspond to that of the Class B1 which had been responsible for hauling that traffic. These adjustments required the fitting of new boilers with a reduced heating surface, a reduced grate area and thicker tyres for the wheels. The tenders were fitted with steam brakes and had their tank capacity increased from 1,800 to 2,176 gallons.

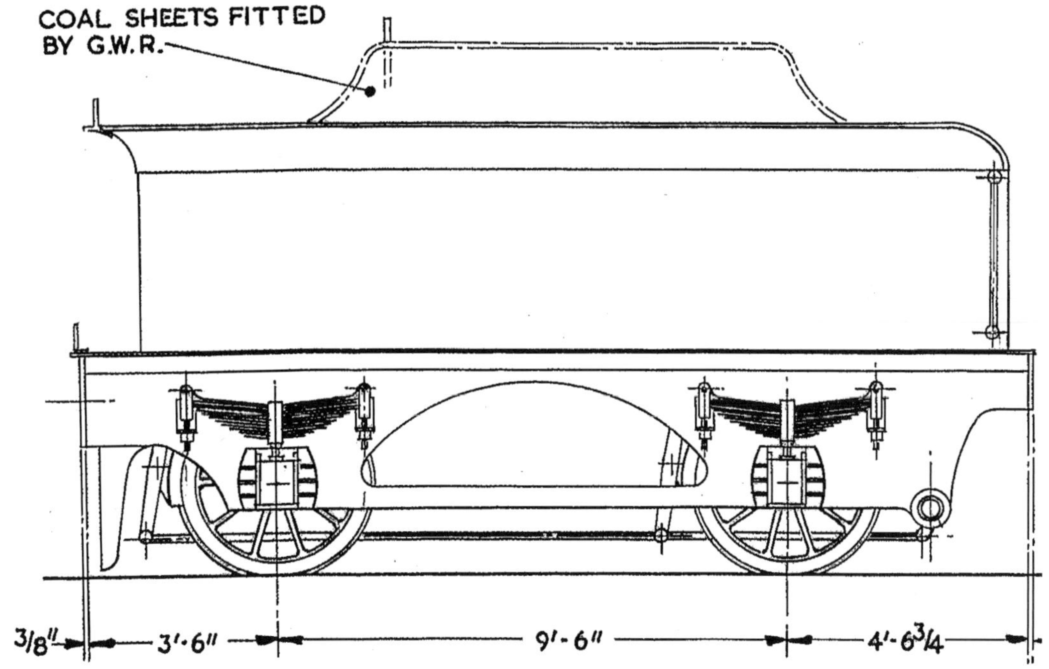

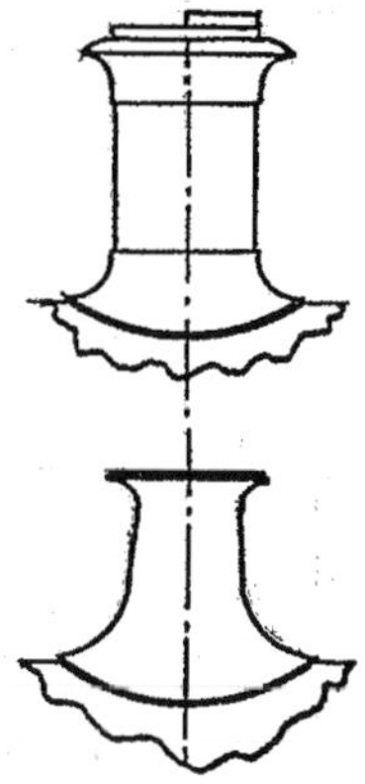

G.W.R. BOILER MOUNTINGS.

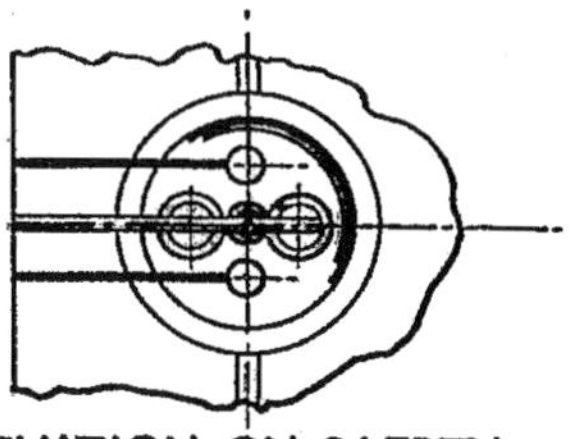

ELEVATION ON SAFETY VALVE SHOWING POSITION OF TWO BELL TYPE WHISTLES.

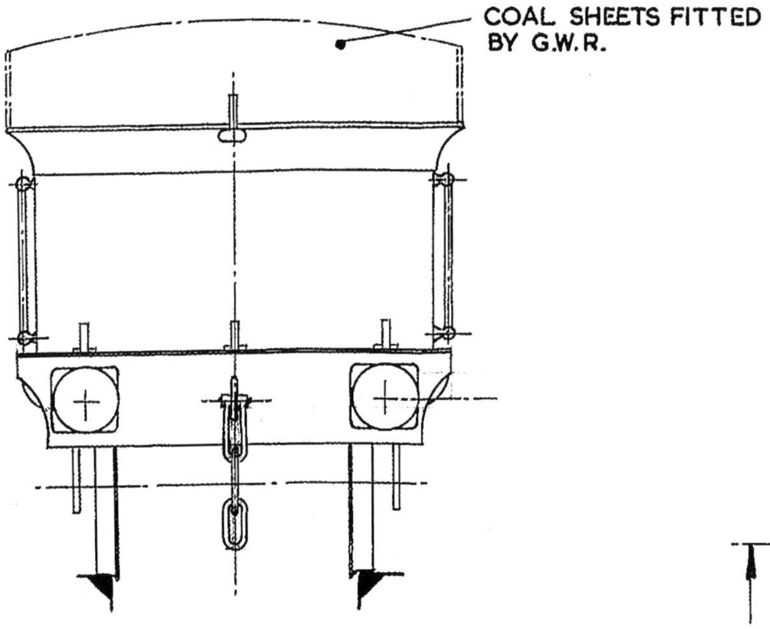

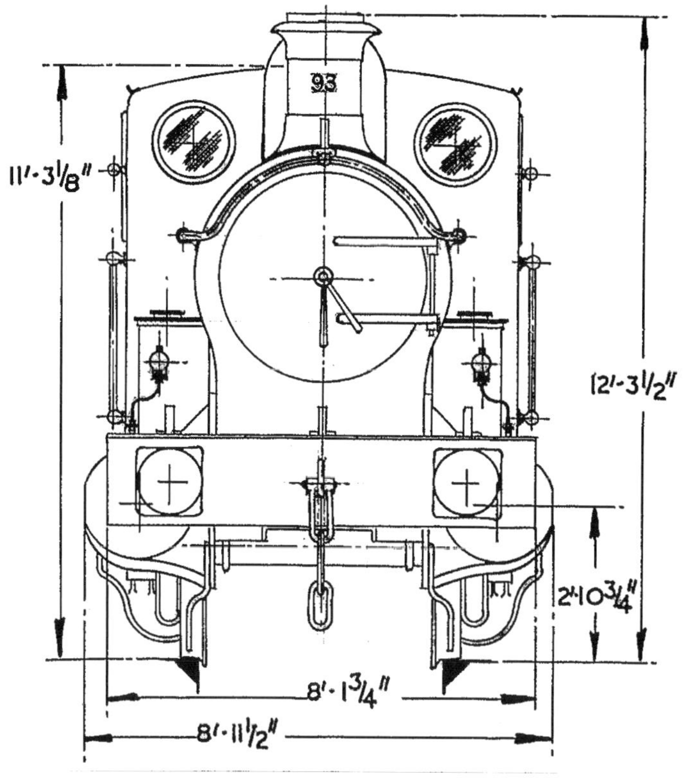

Class D No. 93. This was the only tender engine class used by the Barry Railway and was acquired by them only because the Swedish & Norwegian Railway, which had originally ordered them, defaulted on its payments. This particular locomotive was acquired in March 1897. PHOTOGRAPHER UNKNOWN

Class D No. 92. Both this locomotive and No. 93 were the property of the Swedish Supreme Court for four years while litigation proceeded between Sharp, Stewart and the Swedish & Norwegian Railway over payments. They were eventually released back to Sharp, Stewart in 1895, offered to the Barry Railway the following year and No. 92 was delivered in April 1897. Although the photograph is of an inferior quality, it is the only known photograph showing the Barry Railway livery on the tender. LOCOMOTIVE PUBLISHING CO./REAL PHOTOGRAPHS 54612

Class D No. 93 is pictured here outside Barry shed. All of the Class Ds were initially confined to hauling mineral trains from Cadoxton Yard to the sidings on the high level Barry Docks, from where the coal wagons were taken to the tips. In 1909, Golding decided that Nos. 92 and 93 were to be re-assigned to main line coal duties and they would then often be seen hauling the Brecon & Merthyr traffic. KEN NUNN COLLECTION H1018/LCGB 26 APRIL 1910

CLASS D SPECIFICATION	
Maker	Sharp, Stewart
Numbers	(a) 35, 36, (b) 92, 93
Maker's numbers	(a) 3446, 3447, (b) 3365, 3394
Date built	(a) 1888; (b) 1887
Wheel arrangement	0-8-0
Cylinders	20 in. × 26 in.
Wheels	4 ft 3 in.
Wheelbase	5 ft 3 in. + 5 ft 2 in. + 5 ft 0 in.
Tractive effort	27,200 lb
Tender coal capacity	90 cwt
Tender water capacity	1800 gallons
Weight full	
Locomotive	48 tons 17 cwt
Tender	24 tons 5 cwt

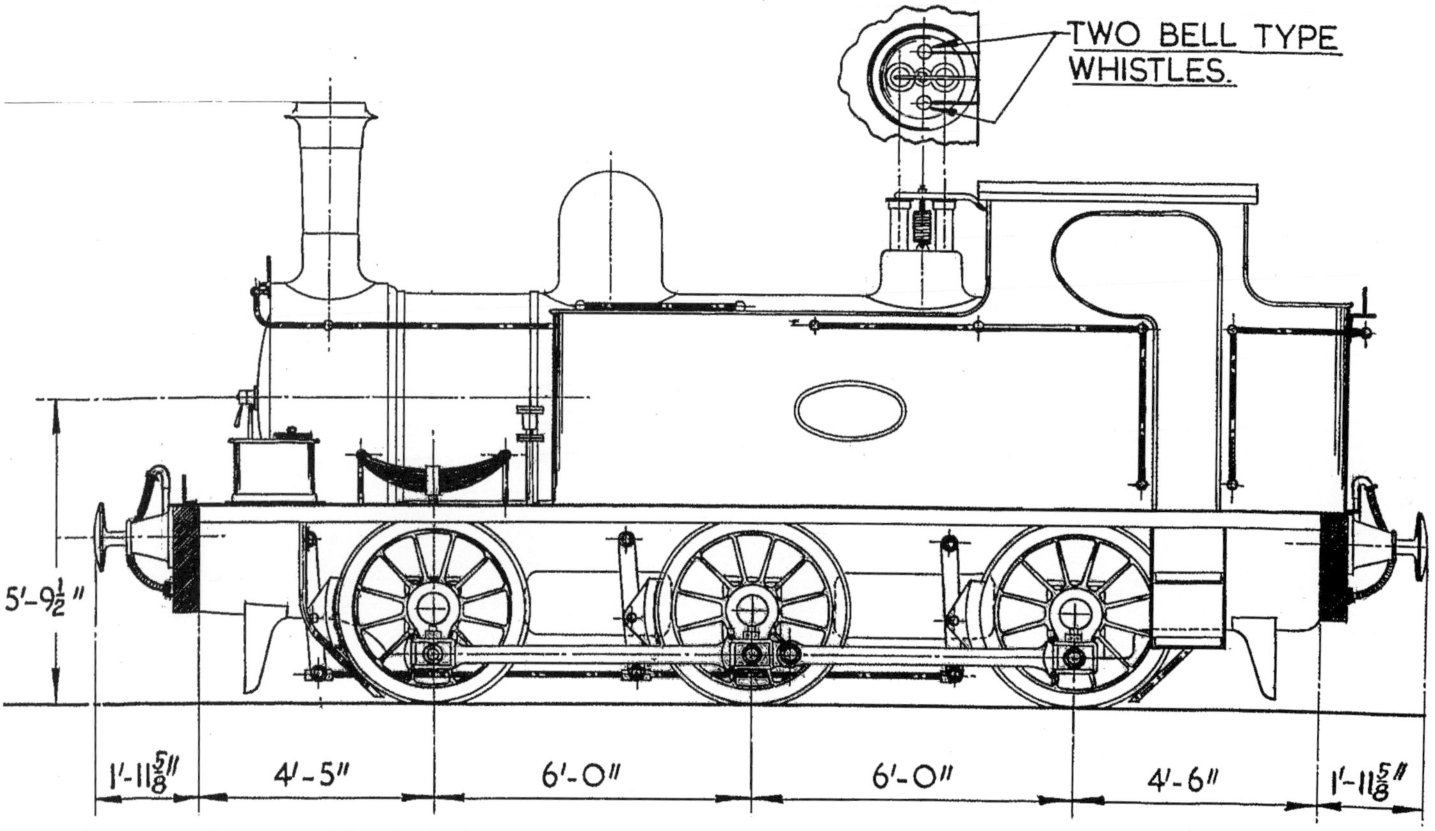

Drawing BRL9: 0-6-0T Class E Locomotive

This class, comprising five locomotives, were delivered by Hudswell Clarke between 1889 and 1891 and were intended to undertake light work in the dock area. The short overall wheelbase of 12 ft made it easier for these locomotives to negotiate the many tight curves to be found at Barry Docks. In addition, it was the only locomotive class to be allowed onto the breakwater opposite Barry Pier station. Their duties were confined to the docks area being used to carry stores and to take coal to the hydraulic power houses.

All five locomotives received new boilers after 1900. In 1909, one of the class, No. 33, was transferred to passenger duties and fitted for auto-working. To accompany this, its wheel arrangement was changed to 0-4-2T by the removal of the rear coupling rods. The two-coach train it pulled became known as the Motor Set and comprised an 1888 four-wheel First and Second Composite coach combined with an 1895 six-wheel Brake Third. The coaches were close-coupled. The service ran between Barry and Llantwit Major.

Originally this ran push–pull with a somewhat elaborate wire-operated mechanism which operated the brakes and whistle but not the regulator. This was controlled by the driver at the coach end giving bell signals to the fireman in the locomotive. However, when Mr Auld took over as Locomotive Superintendent, the service reverted to being loco-hauled with the engine running round at the terminal stations.

Class E No. 50 stands in front of a rake of 6-wheeled coaches and sporting a top headlamp. This locomotive was vacuum-fitted for passenger working but, unlike Nos. 33 and 53, retained its 0-6-0T wheel arrangement. Ken Nunn H1012 of 26 Apr 1910 courtesy LCGB

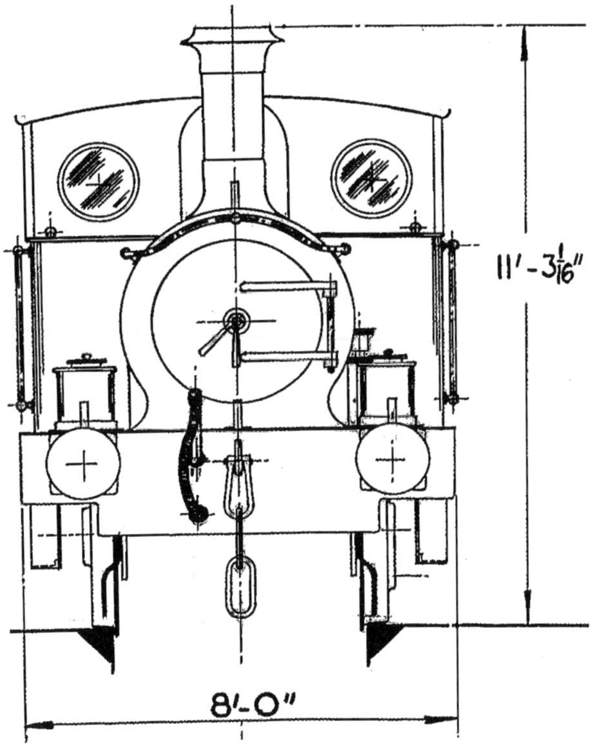

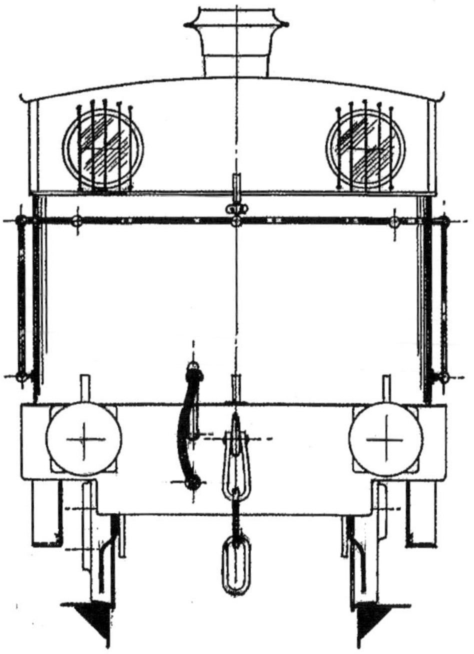

Class E Specification	
Maker	Hudswell, Clarke
Numbers	(a) 33, 34 (b) 50, 51 (c) 53
Maker's numbers	(a) 331, 332 (b) 343, 344 (c) 352
Date built	(a) 1889 (b) 1890 (c) 1891
Wheel arrangement	0-6-0T
Cylinders	14 in. by 20 in.
Wheels	3 ft 3½ in.
Wheelbase	6 ft 0 in. + 6 ft 0 in.
Tractive effort	11,810 lb
Bunker capacity	30 cwt
Tank capacity	660 gallons
Weight	27 tons 10 cwt

Sales to industry of Barry Railway Class E locomotives

GW No.	Former BR No.	Date of sale	Sold to	Location	Scrapped
781	33	04/33	Robert Stephenson & Co. for Hepworth Iron Co Ltd	Hazlehead Bridge	12/60
782	34	11/39	Harton Coal Co Ltd	Boldon Colliery No. 1	03/58
785	53	09/32	Robert Frazer & Sons Ltd for Hartley Main Collieries	Essington	11/58

Class E No. 33. Having arrived at Barry from Llantwit Major and standing in the up platform, No. 33, sporting its 0-4-2T wheel arrangement, is coupled to the Motor Set. The wires used by the driver in the auto trailer to communicate with the fireman in the locomotive can clearly be seen.
LOCOMOTIVE PUBLISHING CO./REAL PHOTOGRAPHS 25001

DRAWING BRL10: 0-6-0T CLASS E LOCOMOTIVE NO. 50 (GWR 783) AS REBUILT BY THE GWR

When the GWR took possession of these engines, certain technical modifications were made to the class. The tube heating surface was reduced from 510 ft^2 to 470 ft^2, the wheels were enlarged from 3 ft 3½ in. to 3 ft 6½ in. and the tractive effort was reduced slightly from 11,810 lb to 11,760 lb. Barry Nos. 33 (GWR No. 781) and 53 (GWR No. 785) had their vacuum brake fittings removed. GWR No. 781 also had GW safety valves fitted. Regarding their GW power and weight classification, the class as a whole was registered as Uncoloured and Ungrouped.

No. 34 (GWR No. 782) was sent to Swindon in 1922 and in March of the following year returned to its duties complete with a GW smokebox and chimney. The weight and weight distribution was increased from 27 tons 10 cwt (9 tons + 9 tons 10 cwt + 9 tons) to 31 tons 16 cwt (10 tons 14 cwt + 10 tons 16 cwt + 10 tons 6 cwt).

In September 1923, No. 50 (GWR No. 783) was also sent to Swindon but it would be another two years and four months before it was seen again, in January 1926. Its modifications were substantial and included a Swindon-designed boiler, a new chimney and a GW coal bunker. It retained its vacuum brake fittings but does not appear to have returned to passenger duties.

The distribution of the class to Barry, Cathays and Cardiff East Dock suggests that they spent their days in shunting duties. Three of the class, Nos. 781, 782 and 785, were sold on (details on page 43) whereas No. 783 and Barry No. 51 (GWR No. 784) survived until nationalisation when they were categorised as British Railways power class 0F.

Renumbering

Barry Railway numbers	Great Western Railway numbers
33, 34	781, 782
50, 51	783, 784
53	785

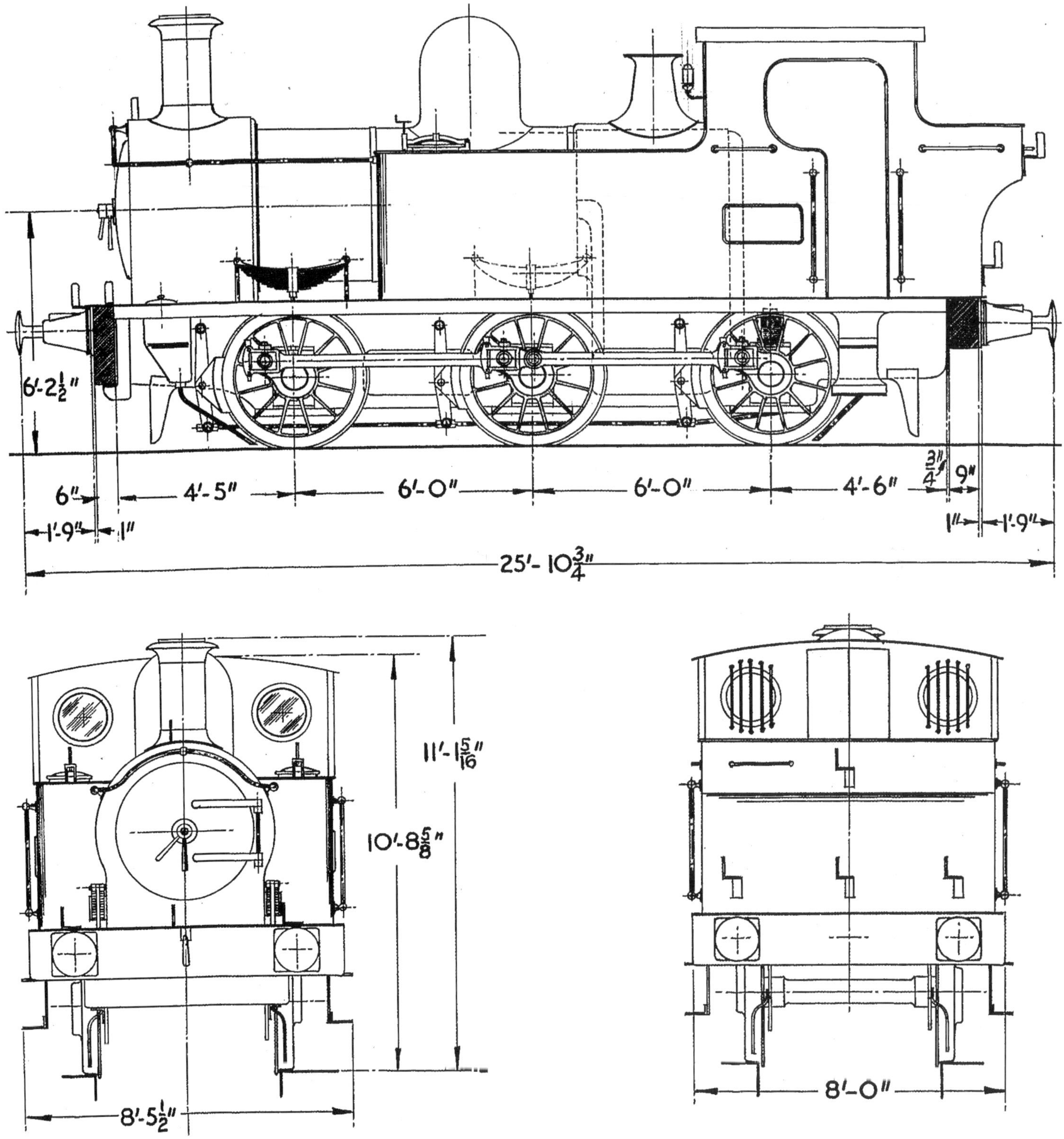
6'-2½"
6"
4'-5"
6'-0"
6'-0"
4'-6"
¾"
9"
1'-9"
1"
1"
1'-9"
25'-10¾"
11'-1 5/16"
10'-8⅝"
8'-5½"
8'-0"

Drawing BRL11: 0-6-0ST Class F Locomotive

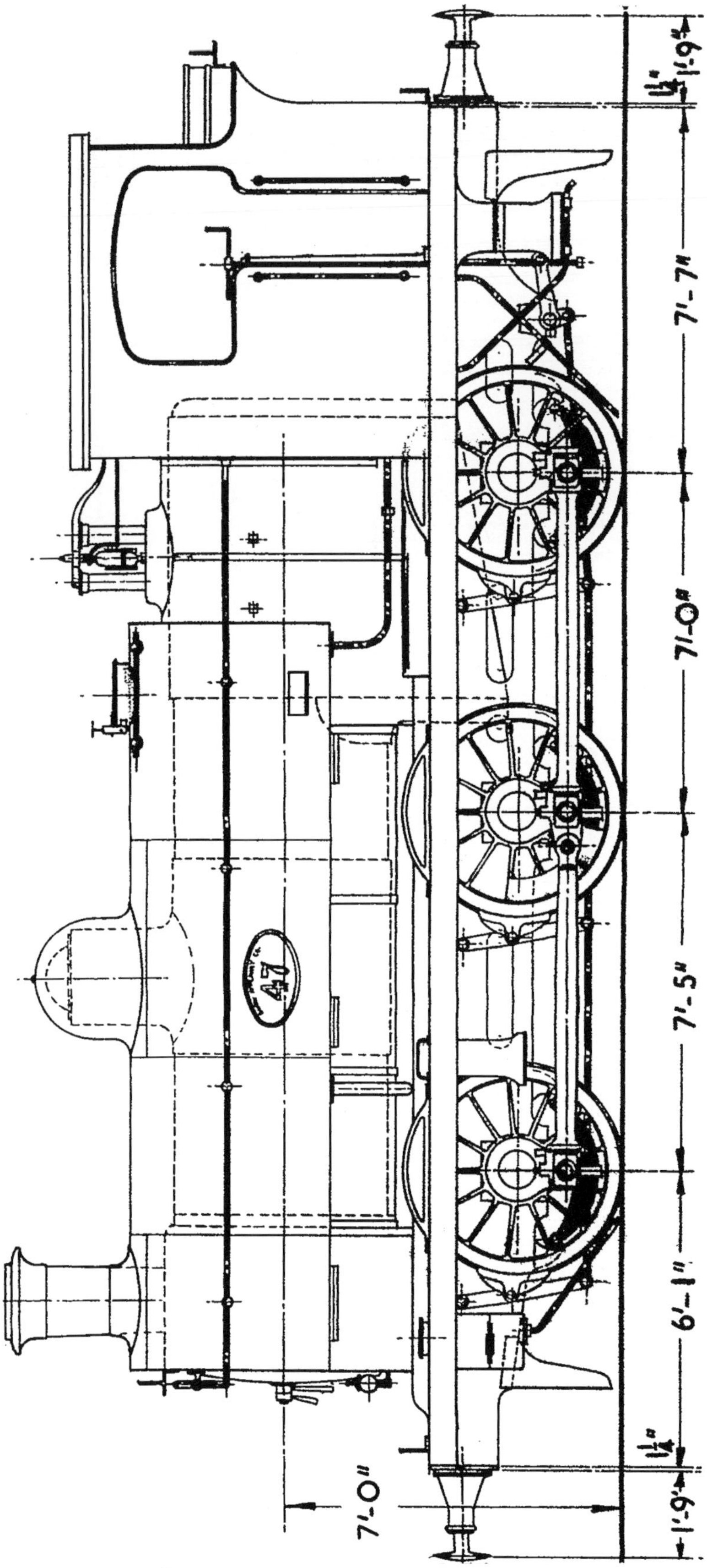

From 1890 onwards, the bulk of the heavy shunting was handled by the Class F, a saddle tank version of the Class A. It was, numerically, the second largest class on the Barry with 28 locomotives. Given the primacy of shunting duties, it is worth noting that lever reverse, usually found on shunting locomotives, did not appear on the Class F until the two 1905 batches, introduced by Hosgood just before his departure. In addition, there is no evidence that lever reverse was fitted retrospectively to the earlier locomotives in the class.

They were built, between 1890 and 1905, by a number of manufacturers: Sharp, Stewart, Vulcan Foundry, North British Locomotive Company and Hudswell, Clarke. They had a bunker capacity of 30 cwt and a tank capacity of 1,050 gallons. The Class F rarely had new boilers fitted as the Locomotive Superintendent at the time, Mr Auld, preferred new boilers to go to main line engines and so this class of locomotives had to make do with repaired boilers.

Although dock shunting was its primary role, the 1920 Barry Railway working timetable listed this class as an alternative to the Class B1 to haul coal trains on the main line, though usually over short distances such as from Peterston Junction to the docks.

It is testament to the effectiveness of this locomotive that so many were sold on in Great Western days. Of the 28, 22 were sold to various collieries mainly in Wales such as those in Bargoed, Treharris and Treorchy. A few went to the North-East of England in the Gateshead area. And of these 22, 20 survived into the era of nationalisation, some longer than others. The last of the class was scrapped in June 1964 at Hafodyrynys Colliery (see page 50).

Renumbering

Barry Railway numbers	Great Western Railway numbers
47–49	710–712
64, 65	714, 715
70–72	716–718
37	708
52	713
99–104	719–724
127, 128	725, 726
129	807
130	729
131	742
132	747
133	754
134–138	776–780

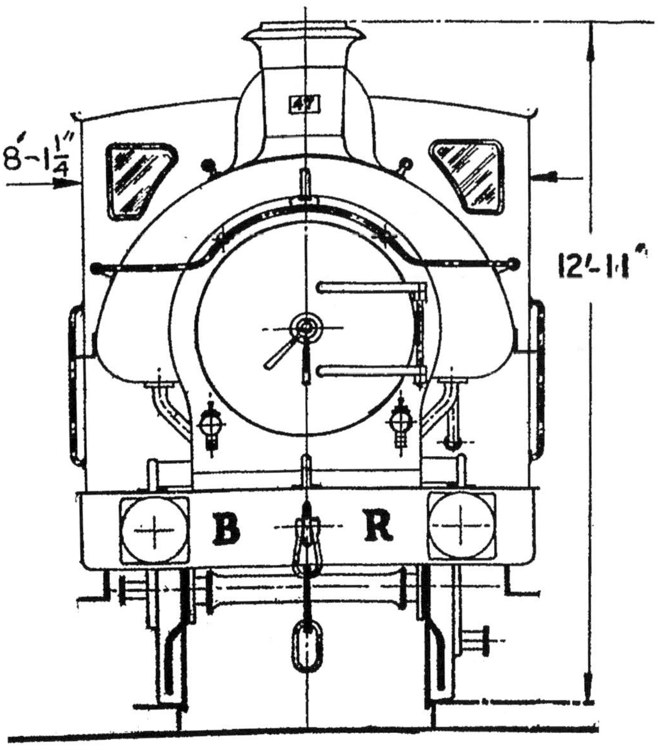

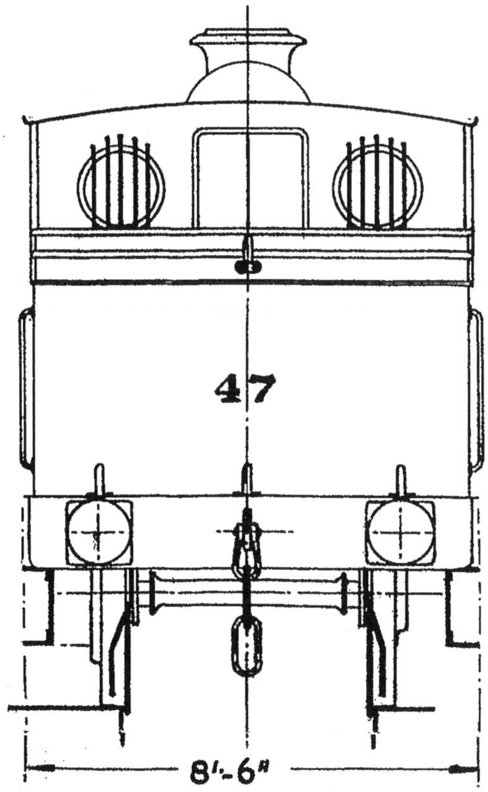

Class F Specification						
Maker	Sharp, Stewart	Vulcan Foundry	Sharp, Stewart	Sharp, Stewart	North British	Hudswell, Clarke
Numbers	47–49	64, 65	70–72	37, 52, 99-104	127–132	133–138
Maker's numbers	3607–3609	1346, 1347	4050–4052	4593–4600	16628–16633	712–717
Date built	1890	1892	1894/5	1900	1905	1905
Wheel arrangement	0-6-0ST					
Cylinders	18 in. × 26 in.					
Wheels	4 ft 3 in.					
Wheelbase	7 ft 5 in. + 7 ft 0 in.					
Tractive effort	20,825 lb					
Bunker capacity	30 cwt					
Tank capacity	1,050 gallons					
Weight full	47 tons 9 cwt	47 tons 9 cwt	47 tons 9 cwt	49 tons 10 cwt	49 tons 10 cwt	49 tons 10 cwt

Resplendent in its manufacturer's grey livery, Class F No. 49 was supplied by Sharp, Stewart in 1890. Altogether four companies built this class of 28: Sharp, Stewart, Vulcan Foundry, North British Locomotive Co and Hudswell, Clarke. They were the workhorses of Barry Docks shunting. SHARP, STEWART & CO

Seen in its natural setting, Class F No. 135 was delivered in 1905 and constructed by Hudswell, Clarke. Their reputation as a class was such that they became very popular as industrial locomotives. 22 of the 28 were sold to various industrial enterprises, including a number of collieries, in Wales and England (see page 50). C W HARRIS COLLECTION

The Class F was the second largest Barry locomotive class to be handed over to the GWR at the Grouping. Prior to that, six Class Fs (Nos. 49, 64, 72, 129, 133 and 137) had been lent to the Great Western in 1917. They worked primarily in Newport, Neath and Duffryn Yard, Port Talbot before being returned.

At the Grouping, the former Class F was given the power and weight classification Blue B. One modification that affected the whole class was the reduction in size of the driving wheels from 4 ft 4 in. to 4 ft 3 in. The class could be divided into two groups: those built between 1890 and 1900; and those built in 1905. The basis for this distinction was that the first group was built with Type 2A boilers whereas the 1905 group was built with Type 2B boilers. The weights of the respective engine groups were also different with the 1890–1900 group weighing 47 tons 9 cwt (15 tons 1 cwt + 16 tons 10 cwt + 15 tons 18 cwt) whereas the 1905 group weighed 49 tons 10 cwt (15 tons 10 cwt + 17 tons + 17 tons). Tractive effort was also greater on the 1905 group. The first group had a 19,525 lb tractive effort whereas the 1905 group tractive effort was 20,825 lb. This was significant when it came to the modifications made by the GWR. The change in wheel size increased the first group's tractive effort to 19,910 lb and the second group's to 21,230 lb.

The class appears to have been sent to Swindon for modification as a series of batches. The first batch (GWR 708, 712, 716, 719, 720, 747, 776 and 778) were fitted with Barry Type 3A boilers with Type 3B tube arrangements or had their original boilers rebuilt to the same specification. The second batch (GWR 710, 714, 718, 721, 723, 725, 780 and 807) received modifications that changed the profile of the locomotive. In addition to the fitting of a Standard No. 9 domed boiler with a single row superheater, pannier tanks replaced the original saddle tank. This increased the tank capacity from 1,050 to 1,200 gallons. A further batch were modified at Caerphilly by having non-superheated Standard No. 9 boilers fitted. By January 1937, all former Class Fs had been withdrawn from Great Western service.

However this was not the end of the locomotives' useful service. They were subsequently sold on to collieries located in South Wales and the North of England, the last of which was withdrawn in 1964, one year short of its 60th birthday.

Drawing BRL12: 0-6-0ST Class F Locomotive as rebuilt by the GWR

Nos. 37, 47, 49, 64, 70, 71, 72, 99, 100, 101, 103, 127, 129, 132, 143, 136 and 138 rebuilt, not all identically.

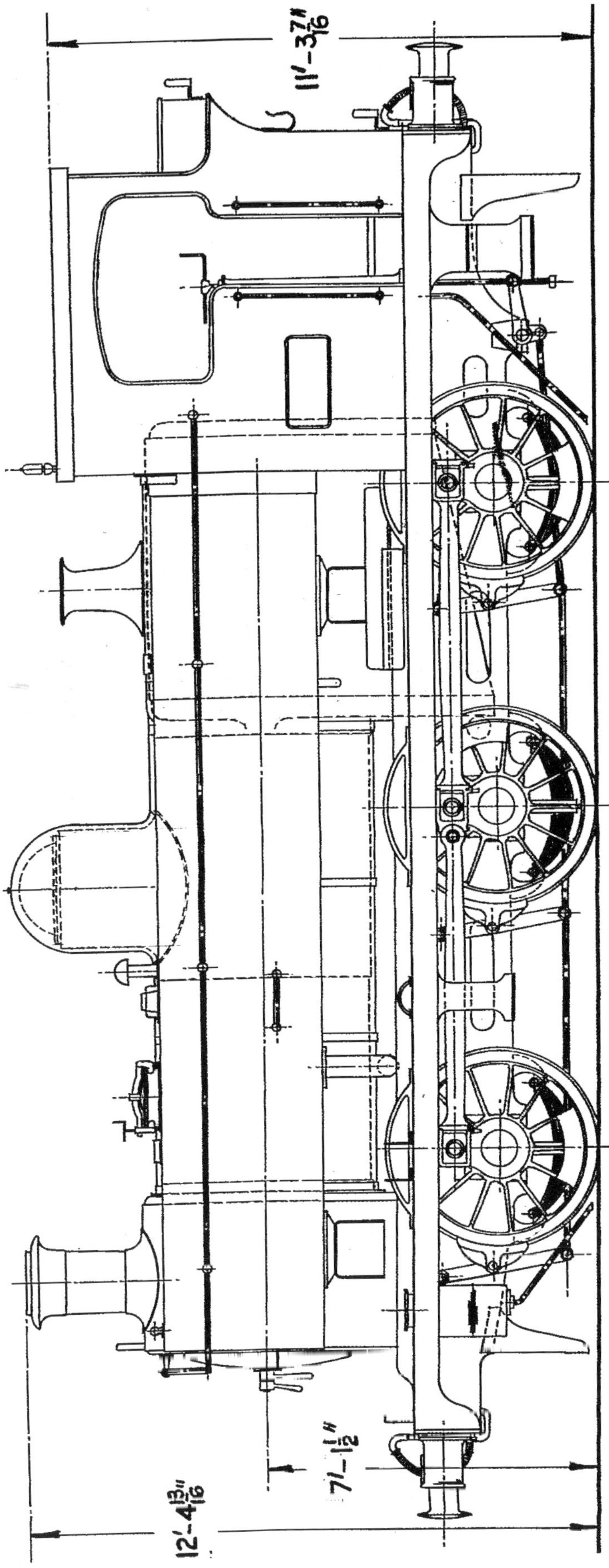

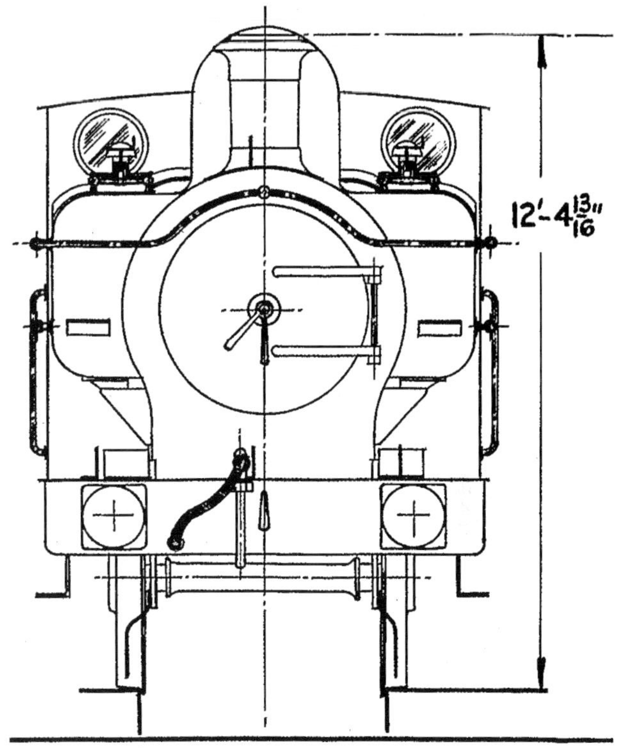

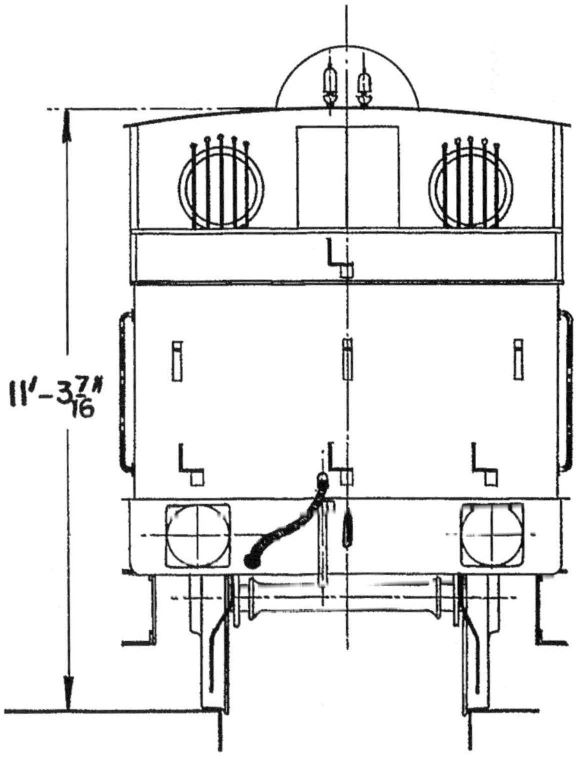

Sales to industry of Barry Railway Class F locomotives

GW No.	Former BR No.	Date of sale	Sold to	Location	Scrapped
708	37	03/33	Powell Duffryn	Bargoed	10/47
711	48	08/34	Powell Duffryn	Bargoed	07/60
712	49	03/33	Coltness Iron	New Mains, Scotland	1959
713	52	07/36	R H Longbotham	Northwood	09/46
714	64	09/34	D S Bowran	Gateshead	1956
715	65	08/32	Ocean Coal Co	Treharris	*c.* 09/50
717	71	11/34	R H Longbotham	Northwood	Not known
718	72	01/35	Ashington Coal	Morpeth	06/62
719	99	09/34	D S Bowran	Gateshead	Not known
720	100	11/36	Partridge Jones & John Paton Ltd	Abersychan	02/61
721	101	05/35	Ocean Coal Co	Treorchy	Not known
722	102	03/36	Robert Frazer	Blyth	*c.* 08/58
723	103	01/35	Ashington Coal	Morpeth	*c.* 10/60
724	104	05/35	Robert Frazer	Hartley, nr Blyth	04/60
725	127	11/34	R H Longbotham	Northwood	02/50
726	128	02/33	R R Paton	Gwauncaegurwen	1957
729	130	09/32	Robert Frazer	Hartley, nr Blyth	Not known
742	131	08/32	Ocean Coal Co	Treorchy	11/62
747	132	05/33	Robert Frazer	Hartley, nr Blyth	Not known
754	133	01/37	Powell Duffryn	Bargoed	*c.* 07/60
776	134	02/33	Powell Duffryn	Aberaman	04/60
780	138	05/36	Burnyeat Brown	Hafodyrynys	06/64

Looking remarkably clean, Class F No. 37, another Sharp, Stewart locomotive, stands in Cadoxton Yard complete with enginemen. C W HARRIS COLLECTION

The Barry Railway undertook a great deal of engineering work in the construction of its network, one of the most impressive examples being Pwll-y-Pant viaduct (or Llanbradach Viaduct as it was also known). Completed by 1905, it was 800 yards long with 11 spans of steel lattice girders balanced on ten piers, eight of which were over 100 ft high. It was demolished by the Great Western Railway in 1937. FROM A CONTEMPORARY POSTCARD

Walnut Tree Viaduct was the first of the major viaducts to be built on the Brecon & Merthyr extension from Tyn-y-Caeau Junction. It was built in 1901 and was 516 yards in length with a maximum height of 120 ft. Given that the narrowing of the Nantgarw Gap at this point created a wind tunnel effect, the viaduct was constructed of steel lattice girders on top of brick built piers. It was demolished by British Railways in 1969. FROM A CONTEMPORARY POSTCARD

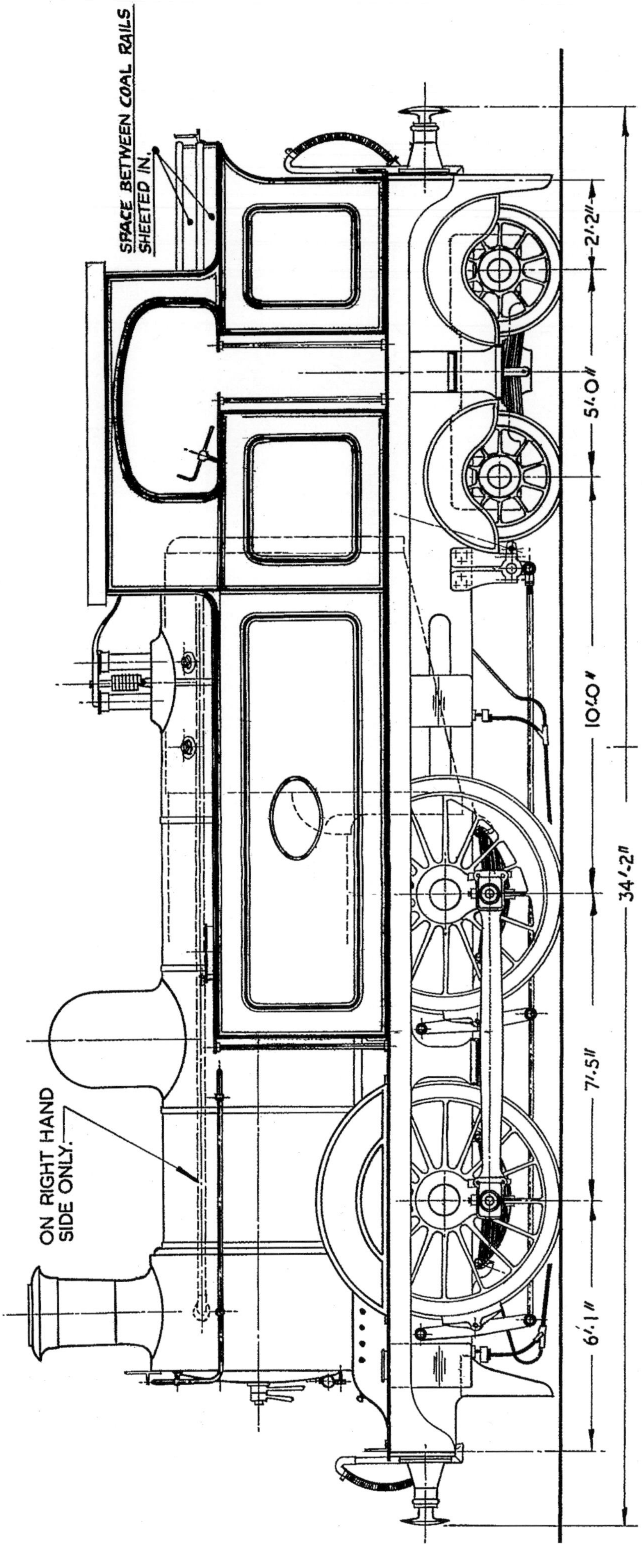

Drawing BRL13: 0-4-4T Class G Locomotive

This was the second of the Barry Railway's passenger locomotive classes after the Class C. It was superior to its predecessor in many ways, two of which were its speed and quick acceleration, an ideal combination for a locomotive working commuter traffic. Two were introduced in 1892 with a further two in 1895. When the first batch was ordered, the only passenger service in operation was between Barry and Cogan. The service was extended to Cardiff Riverside in August 1893 and in April 1894 to Cardiff Clarence Road. The second batch of March 1895 would undoubtedly have helped with this extended service as well as the passenger service between Barry and Porth which began in March 1896.

It was the policy of the Barry Railway for its passenger locomotives to be immaculately presented, and part of the original design was the provision of ornamental splashers over the trailing bogies. These, however, were removed by Mr Golding in 1910. At the same time, Golding introduced his customary modifications to the locomotive, namely thicker tyres, sheets around the coal rails and the reinstatement of slide valves.

A change in duties came about in 1897 with the delivery of the last Barry passenger class, the Class J, which took over the Barry to Cardiff service. The Class G transferred to hauling passenger trains on the main line from Barry to Porth as well as the new Cardiff to Pontypridd service via St Fagans which began in June 1897.

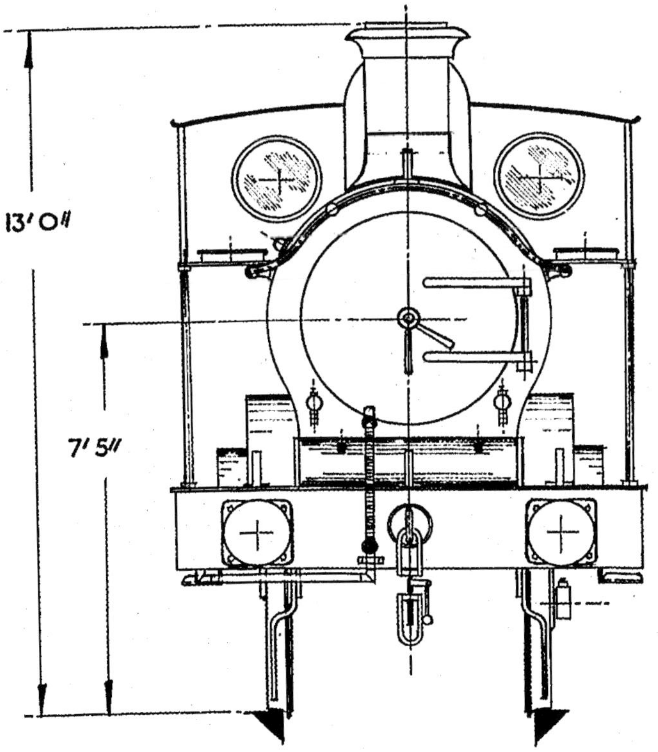

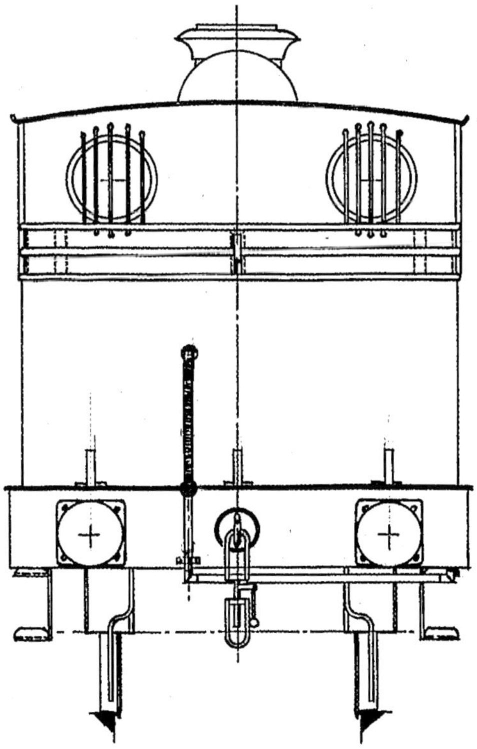

Class G Specification		
Maker	Vulcan Foundry	Sharp, Stewart
Numbers	66, 67	68, 69
Maker's numbers	1348, 1349	4053, 4054
Date built	8/1892	3/1895
Wheel arrangement	0-4-4T	
Cylinders	18 in. × 26 in.	
Wheels		
Coupled	5 ft 7½ in.	
Trailing	3 ft 0 in.	
Wheelbase	7 ft 5 in. + 10 ft 0 in. + 5 ft 0 in.	
Tractive effort	15,925 lb	
Bunker capacity	30 cwt	
Tank capacity	1,400 gallons	
Weight full	56 tons 1 cwt	

Class G No. 68 is seen here on passenger duty. It replaced the previous passenger Class C when introduced in 1892. There were only four of this class and this particular locomotive was constructed by Sharp, Stewart and delivered in March 1895. CLIFFORD COLLECTION NO. 28

Class G No. 67 was one of the first batch of this class to be delivered in August 1892 from Vulcan Foundry. In August 1902, Hosgood experimented with this particular locomotive by replacing the slide valves with piston valves. The experiment lasted for eight years before the slide valves were reinstated. CLIFFORD COLLECTION NO. 27

Class G No. 67 is lined out in Vulcan Foundry photographic grey livery. This class was fitted with a Type 2A boiler and was renowned for its fast acceleration which was important for the commuter work it undertook between Barry and Cardiff. Being classed as passenger engines, the locomotives were fitted with polished domes and safety valve bases as well as brass chimney caps. VULCAN FOUNDRY

Painting specification for the Class G locomotives, taken from the Barry company's "Invitation to tender"
Before any paint is applied the whole surface of iron must be thoroughly cleansed from rust and scale. The boiler is to receive two coats of red oxide paint before being lagged. The whole of the boiler, lagging plates, frames, buffer beams, wheels, side tanks, coal bunks, and cab to have one coat of lead color, two of filling up, and stopped with hard stopping, and rubbed down to a perfectly smooth surface, then painted with two coats of lead colour, each faced down with pummice stone, and afterwards three coats of chocolate, to sample. The smoke box, ash pan, back of firebox, foot plate, brake work, springs, &c., to have one coat of lead color paint and two coats of Japan black. (*sic*)

The inside of frames and cross stays to have two coats of red oxide paint. Inside of cab to be painted to sample.

The axles to have two coats of white lead, and two coats of black paint.

The painting on outside of the Engine to have three coats, and inside one coat of best hard drying engine copal varnish, as supplied by Messrs Tabo & Trego to this Company.

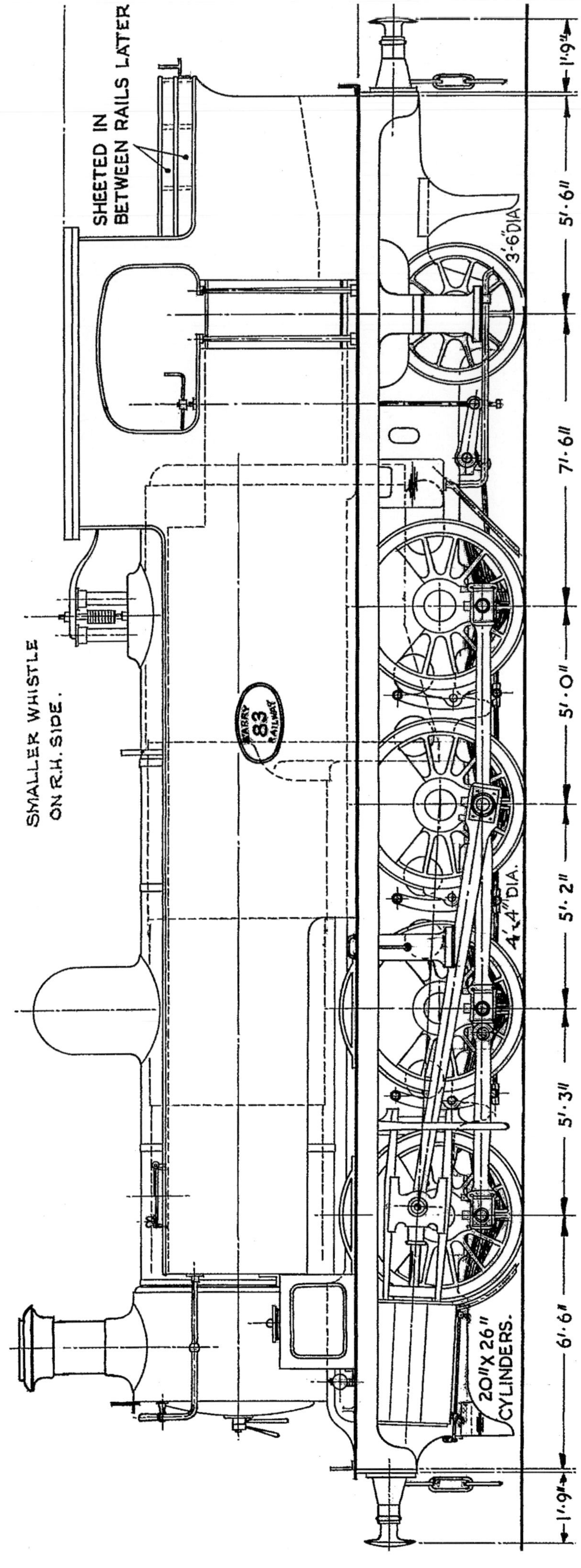

Drawing BRL14: 0-8-2T Class H Locomotive

Built in 1896 by Sharp, Stewart, these locomotives were intended for hauling heavy coal traffic over the Vale of Glamorgan Railway between Bridgend and Barry. However, this line did not open until December 1897 and so this class was assigned to mineral traffic from both Porth and Treforest over the main line to the docks. It was effectively a tank engine version of the Class D, having the same tractive effort at 27,200 lb; these two classes of locomotives were the most powerful on the Barry Railway.

As it turned out, the less powerful B1 – tractive effort 21,060 lb – proved more than adequate for hauling mineral trains over the Vale of Glamorgan Railway thus rendering the Class H redundant for that role. Consequently, they were given the task of taking coal trains from Cadoxton Yard to the dock's tip sidings. Because of the power of the engine, they were able to haul trains of around 100 coal wagons.

Like other locomotives in the company, the Class H was modified to include thicker tyres and the enclosure of the coal rails. In 1910, John Auld designed a new boiler used on both the Class D and Class H, the consequence being that the boiler pressure was increased to 160 lb/in^2.

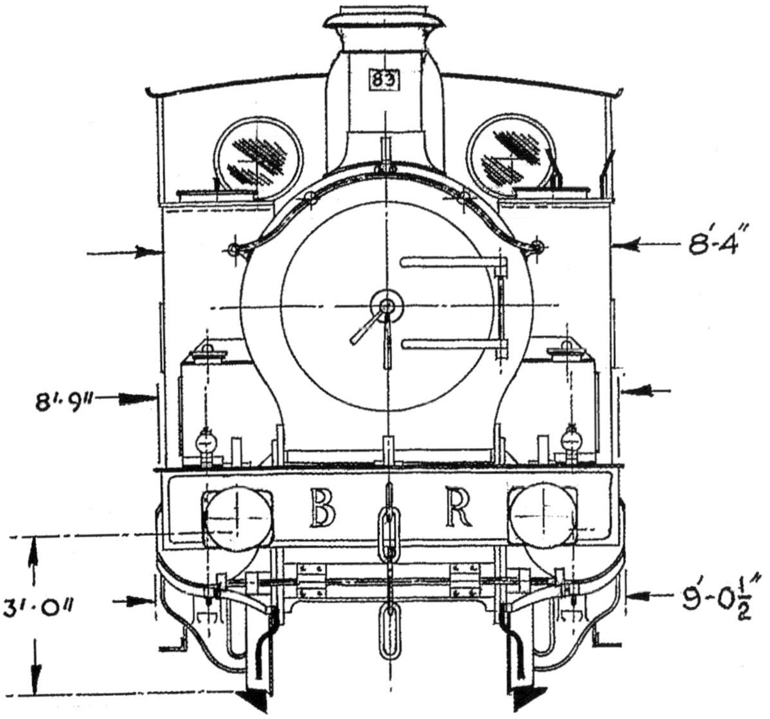

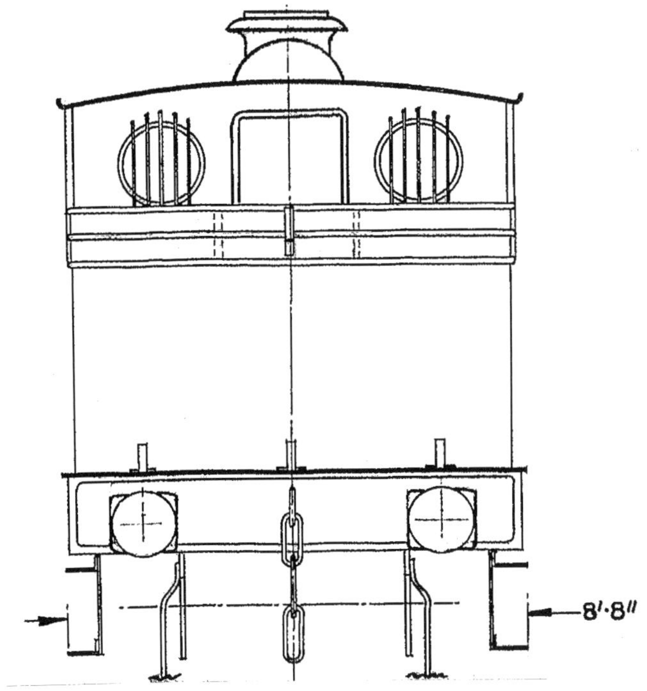

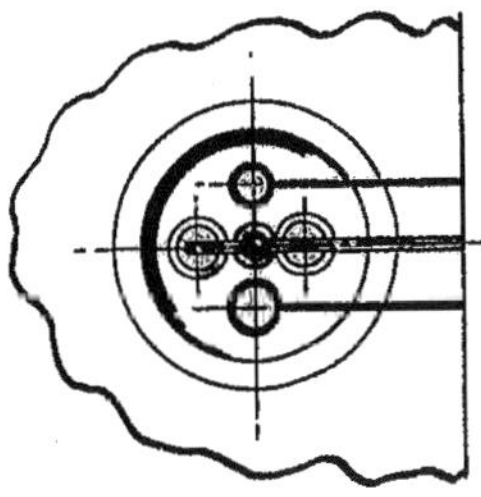

PLAN ON SAFETY VALVE SHOWING POSITION OF TWO BELL TYPE WHISTLES.

Class H Specification	
Maker	Sharp, Stewart
Numbers	79–85
Maker's numbers	4182–4188
Date built	1896
Wheel arrangement	0-8-2T
Cylinders	20 in. × 26 in.
Wheels	
Coupled	4 ft 3½ in.
Trailing	3 ft 6 in.
Wheelbase	5 ft 3 in. + 5 ft 2 in. + 5 ft 0 in. + 7 ft 6 in.
Tractive effort	27,200 lb
Bunker capacity	70 cwt
Tank capacity	2,100 gallons
Weight	74 tons 16 cwt

Class H No. 84. These 0-8-2Ts were a development of the Class D 0-8-0s; together they were the most powerful locomotives on the system. They were built by Sharp, Stewart and delivered in 1896. The design was used as a basis for three locomotives ordered for the Port Talbot Railway from the same company in 1901.
C W Harris collection

Class H No. 79, seen in its natural habitat, performs the duty with which it was synonymous, marshalling wagons in Cadoxton Yard and hauling coal trains to Barry Docks sidings. Cope 364/R Carpenter collection 24 July 1922

The same locomotive is shown at rest and looking remarkably clean. These locomotives were known occasionally to venture out onto the Brecon & Merthyr Extension as far as Penrhos Junction and also to Peterston Junction on the GWR. Originally, the second set of wheels were fitted as flangeless. However the wear on the other wheels caused the second set to bind between rails. Thin flanges were subsequently added to the second and third set together with special tyres to help prevent binding. Ken Nunn collection H1015/LCGB 26 April 1910

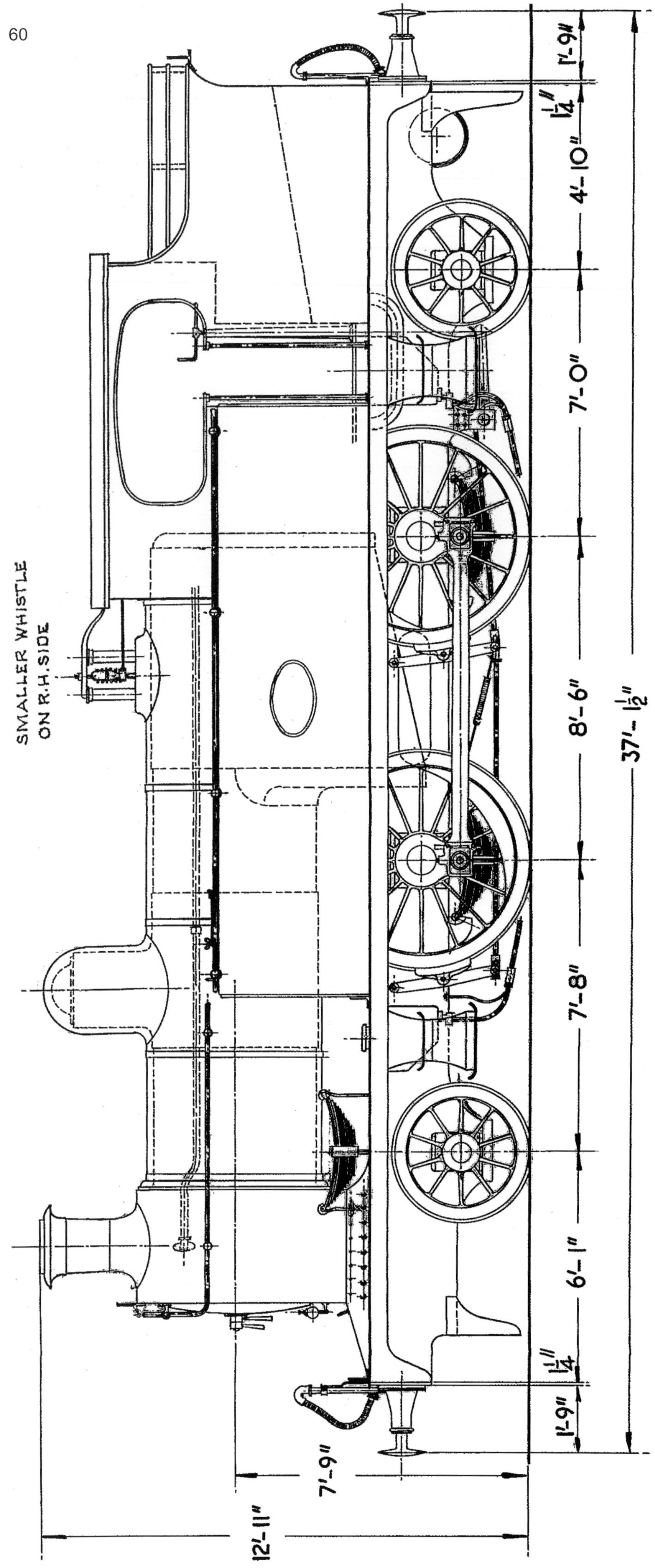

Drawing BRL15: 2-4-2T Class J Locomotive

This was the last of the passenger class engines designed for the Barry Railway and, although a 2-4-2T, used many of the same components from its predecessor, the Class G. There were two manufacturers of the Class J: the first three coming from Hudswell, Clarke and delivered in 1897 and the final eight coming from Sharp, Stewart between 1898 and 1899. Originally the Hudswell, Clarke locomotives were known as the Class J and the Sharp, Stewart as the Class J1. There were some differences between the two designs including tank capacities, the mainframes, the sanding mechanisms and the cab shape but nothing that made them readily distinguishable from each other.

In traffic, they were to be seen hauling passenger trains on the Barry to Cardiff service, or the Cardiff Branch, as the company liked to call it. But with the Barry operating three passenger services on its network (not including the Vale of Glamorgan services), Barry to Clarence Road, Barry to Porth and Clarence Road to Pontypridd, 15 engines from both the Class G and Class J were hard pushed to cover the service, especially if a locomotive was in for repair. In these circumstances, Class B1s were drafted in to fill the gap.

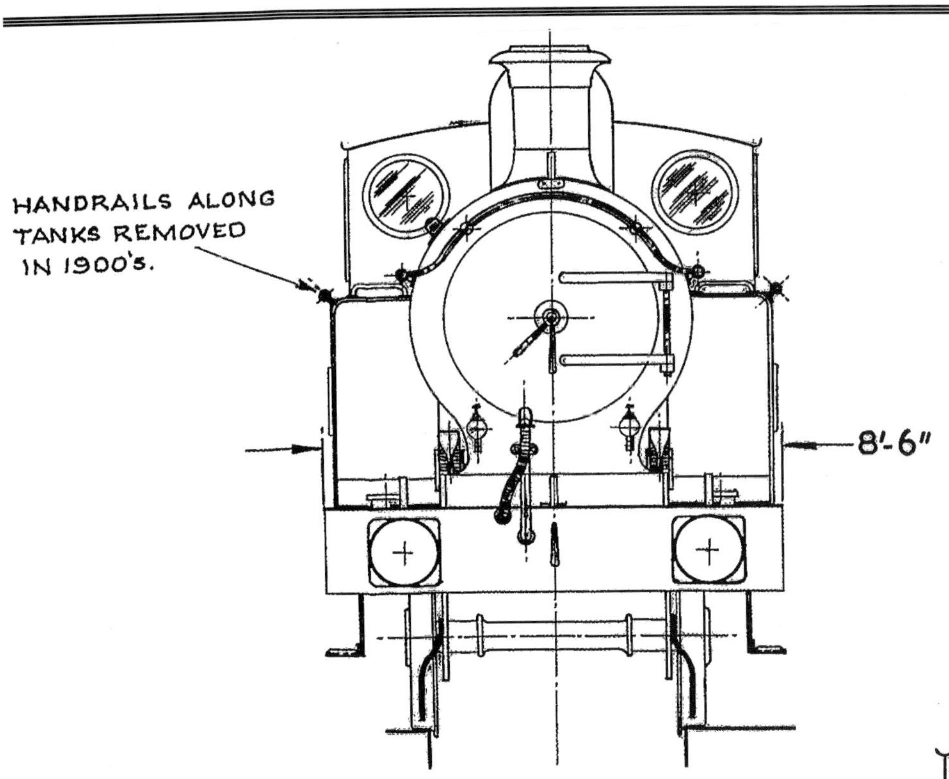

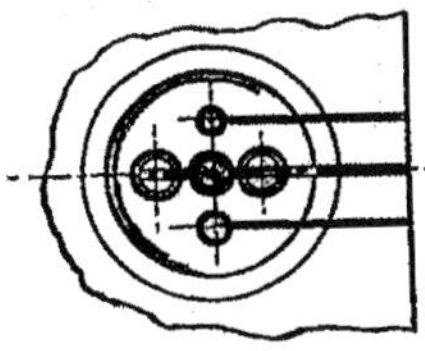

PLAN ON SAFETY
VALVE SHOWING
POSITION OF TWO
BELL TYPE WHISTLES.

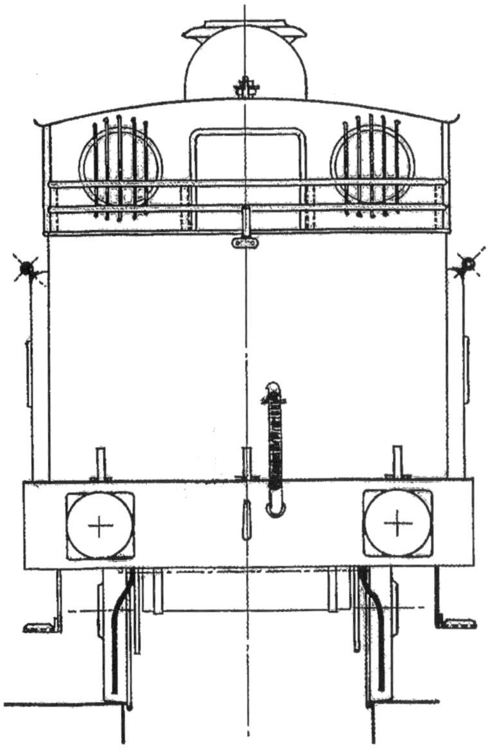

Class J No. 90. 11 of this class were constructed by both Hudswell, Clarke and Sharp, Stewart. The locomotive in the photograph is an example of the first batch delivered from Sharp, Stewart in 1898. Initially the Hudswell, Clarke engines were categorised as Class "J" and the Sharp, Stewart engines as Class "J1". Locomotive Publishing Co/Real Photographs 54609

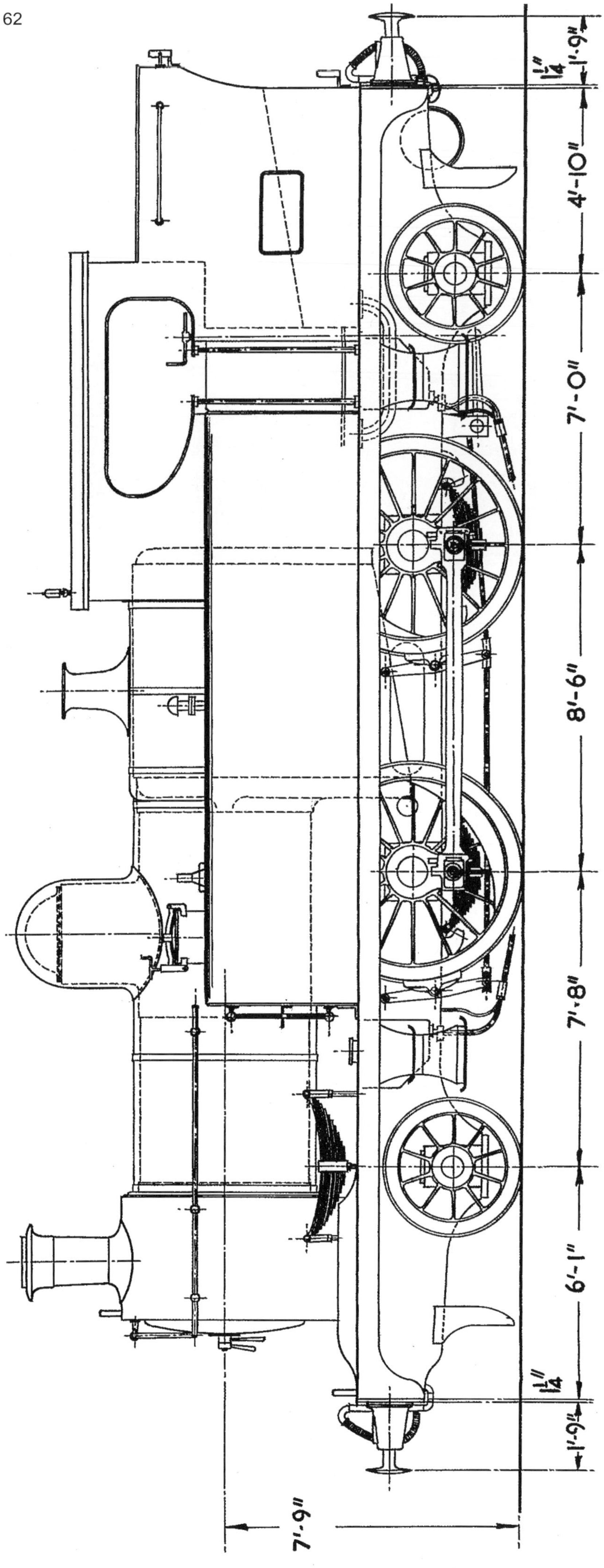

Drawing BRL16: 2-4-2T Class J Locomotive as rebuilt by the GWR

Prior to the GWR taking ownership of all Barry Railway locomotives on 1st January 1923, four Class Js (Barry Nos. 90, 95, 96 and 97) went to Swindon in November 1922 to be examined and, if required, modified. Their boilers were rebuilt to GW specifications, new smokeboxes were fitted with GW chimneys and rebuilt bunkers. They returned to traffic duties in 1924.

Later, No. 87 (GWR No. 1312) was sent to Swindon for a rebuild. This included a Standard No. 9 domed boiler, a single-row superheater, a standard bunker and weight redistribution of its 56 tons 9 cwt (13t 6c + 16t 14c + 16t 14c + 9t 15c). It returned to traffic duties in September 1926. Yet in the same year, four of the class, which had not been through the modification or rebuild process, were condemned.

These rebuilt locomotives were shedded at Barry but it would not be long before the Great Western 2-6-2T 3901 class replaced the Class J. The last of the class was withdrawn in July 1930 and were cut up in Swindon by July 1932.

Renumbering

Barry Railway numbers	Great Western Railway numbers
86–91	1311–1316
94–98	1317–1321

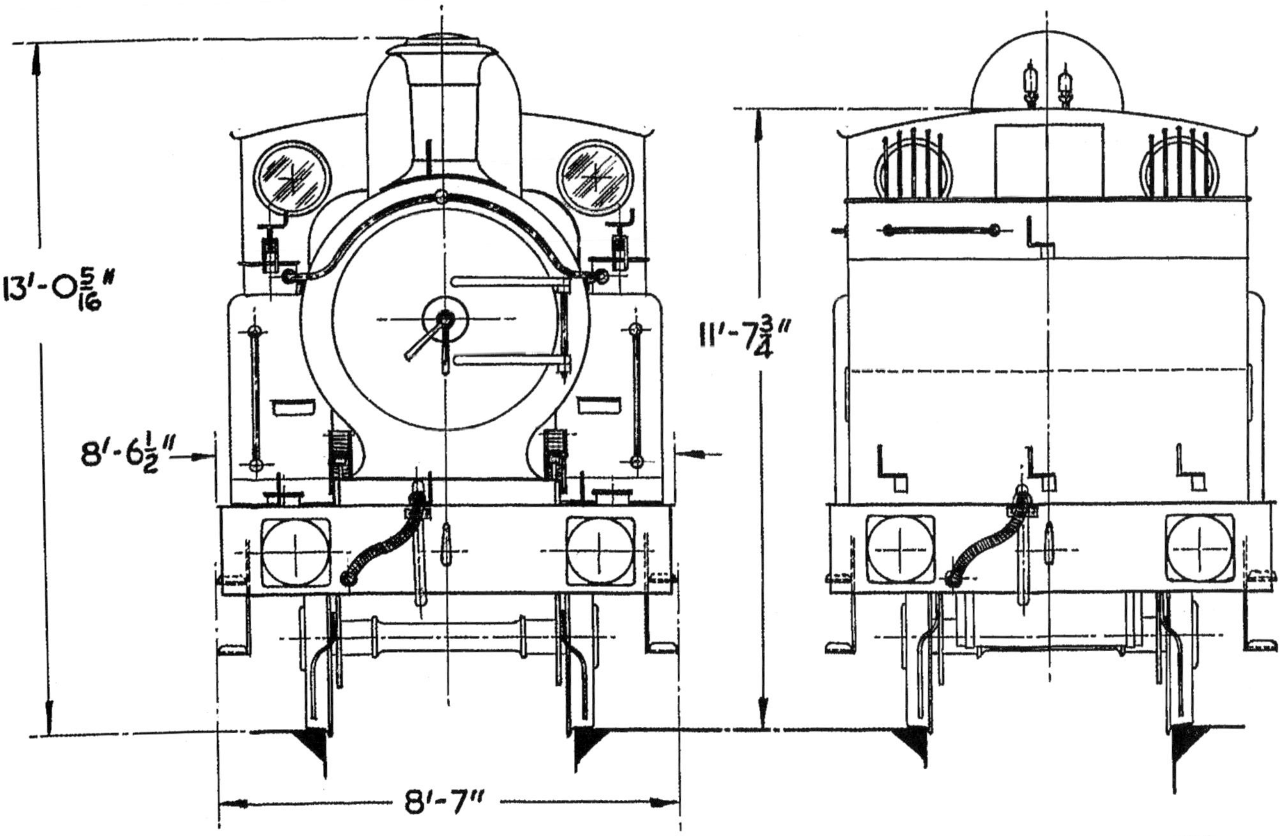

Class J Specification			
Maker	Hudswell, Clarke	Sharp, Stewart	Sharp, Stewart
Numbers	86–88	89–91	94–98
Maker's numbers	473–475	4367–4369	4497–4501
Date built	1897	1898	1899
Wheel arrangement	2-4-2T		
Cylinders	18 in. × 26 in.		
Wheels			
Coupled	5 ft 7½ in.		
Carrying	3 ft 6 in.		
Wheelbase	7 ft 8 in. + 8 ft 6 in. + 7 ft 0 in.		
Tractive effort	14,930 lb at 150 lb/in^2		
Bunker capacity	70 cwt		
Tank capacity	1600 gallons	1550 gallons	1550 gallons
Weight full	60 tons 18 cwt		

A Hudswell, Clarke Class J No. 86 heads a train entering Cardiff Riverside having started at Cardiff Clarence Road with a service for Barry. They had the usual polished chimney caps, dome covers and safety valve bases of all of the Barry's passenger locomotives. It was rumoured that Golding had intended to fit speed recorders to this class but there is no evidence of this having taken place. KEN NUNN COLLECTION 1551/LCGB 11 AUGUST 1913

Class J No. 88. This locomotive was from the Hudswell, Clarke batch. It differed slightly from the other locomotives in that, together with No. 97, it later received a copper chimney cap as opposed to the usual brass variety. This was the last passenger locomotive class built for the Barry.
KEN NUNN COLLECTION H1017/LCGB 26 APRIL 1910

Barry Island Signalbox. A typical Barry Railway signalbox, owing a lot to the work of the contractor, Evans O'Donnell, for the Great Eastern Railway, notably the fenestration with six small panes at the top of the windows, although the design of the bargeboards is unique to the Barry. Believed opened during 1896, it started life with a 22 lever frame, replaced by one of 27 levers in 1903, and yet again by one of 45 levers in 1920, indicative of the expanding excursion traffic. It was closed in 1929, when the GWR provided East and West boxes to cater for the still expanding demand. LENS OF SUTTON

Opening of the Barry Island Railway

(Extract from the *Barry Dock News*, Friday August 7th 1896)

"The first station agent on Barry Island is Mr Daniel Fisher, late signalman at Barry; and the first passenger train reached Barry Island station on Monday morning shortly after 9.30, bringing therewith hundreds of passengers, principally holiday seekers. A large number of fog-signals were discharged by the engine along the line between Barry station and the Island, and the run across the harbour was much admired. The train was in charge of Mr O.T. Rees as guard, and was driven by Mr. F. Burgess, and amongst the gentlemen present to welcome the arrival of the train at the new destination were Mr. R. Evans (general manager of the Barry Company), Mr. H. J. Vincent (assistant manager), Mr. J. Bell, C.E. (resident engineer), Mr. J.H. Hosgood (locomotive superintendent), and Inspector Butler, the general public being represented by Alderman the Rev Aaron Davies (Pontlottyn), Alderman J.C. Meggitt, J.P., Councillors W. Thomas and D. Morgan, Mr B. Lewis, Mr H. Chappell, Mr D. Sibbering Jones, Mr J.C. Pardoe, A.M.I.C.E. (surveyor), &c. Mr F. T. Rednall, contractor's agent, was also present. The first departure train left at 10.2, an indication of the complement of passengers being afforded by the fact that the *Barry Dock News* representative in attendance was supplied at the booking-office with ticket No. 800. About a score of fully-laden trains ran to the Island during the day, and it is computed that the passenger traffic by rail to the Barry district on Monday could not have been far short of 30,000. We should add that Mr J.H. Insole, J.P., one of the directors of the Barry Co., also visited Barry Island on Monday, and, with Mr Robert Forrest, J.P., also a director, was highly pleased with the arrangements.

The Barry Company have decided to issue cheap return tickets between Cardiff and Barry Island every Wednesday and Saturday afternoon during the summer months."

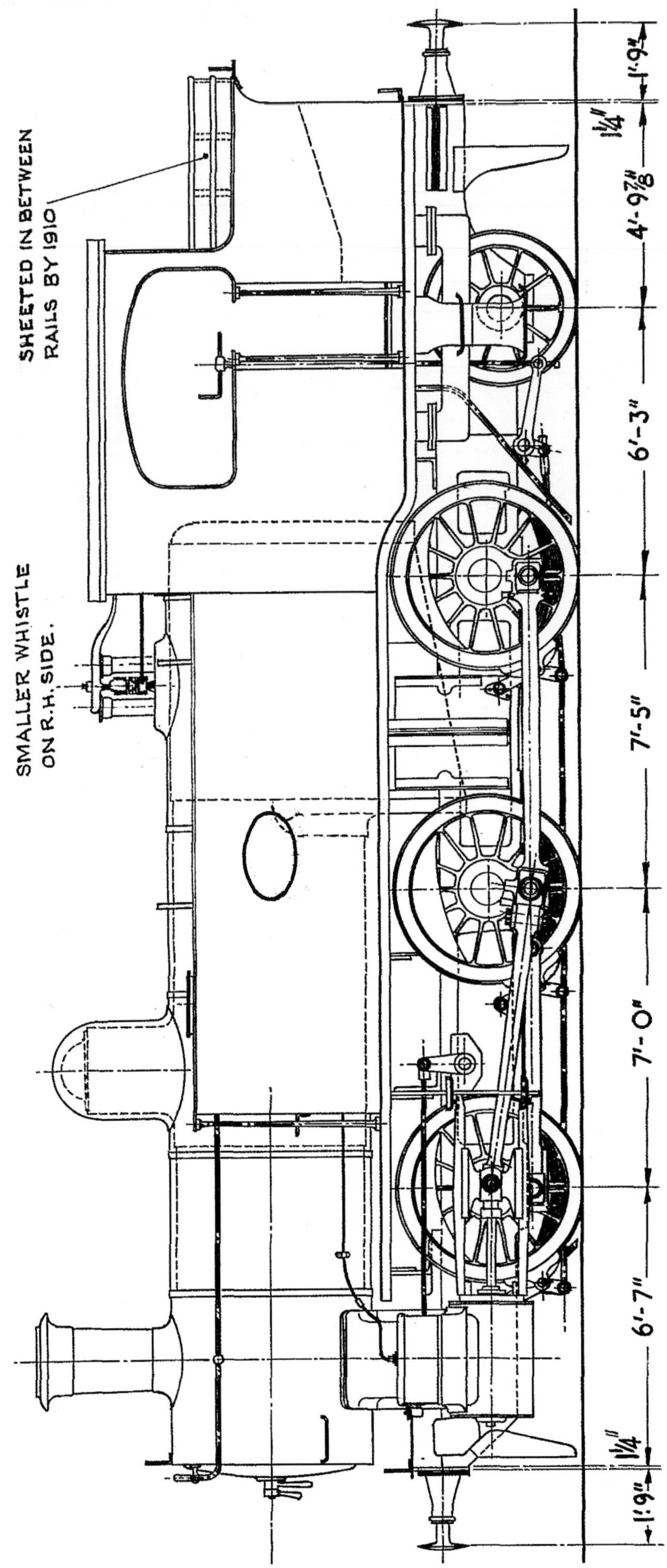

Drawing BRL17: 0-6-2T Class K Locomotive

In 1899, the Barry Railway was in need of more locomotives to perform the same duties as the Class B1. However, no locomotive manufacturer in the UK was available to deliver them as all their order books were full. This meant the Barry had to look elsewhere and settled on the Cooke Locomotive & Machine Company in Paterson, New Jersey, about 20 miles north west of New York. It was intended that the locomotives should generally conform to the dimensions of the Class B1. However, to conform strictly to this would have a meant a delay in the delivery of the locomotive and so certain compromises were reached.

The locomotive was described as having a Barry back end and an American front. Such features substantiating such a claim were outside cylinders, top-mounted steam chests and slide valves using a long valve spindle.

Interestingly, the locomotives were not delivered complete but in parts and thus needing to be erected; this was done at Barry Works on the dock site. Unsurprisingly, they were known to the enginemen and other railway staff as "Yankees". They were as powerful as the B1, but in service the locomotives proved to be heavy users of coal and water. As a former Barry driver once said, "You dare not pass a single water column with the Yankees." Given that the *raison d'etre* of the B1s had been their ability to haul a train from Barry to Porth without the need for a stop, this deficiency only served to undermine their usefulness.

Attempts were made to reduce steam consumption by reducing the size of the steam ports but this modification did not prove a success and so the Class K was taken off main line mineral traffic. As a result, two of the locomotives were sent to Hafod shed to help with banking duties from Trehafod to Pontypridd and from Treforest Junction to Tonteg. The other three were used for hauling wagons from Cadoxton Yard to Barry sidings. Subsequently two of these were allocated to Hafod, again for banking duties and the fifth remained in Barry as the shed pilot.

One special duty given to the Class K was the annual Good Friday trip from Trehafod to Barry for company staff and their families to attend an orchestral concert. The Class K was used primarily because they were the only vacuum-fitted locos allocated to Hafod.

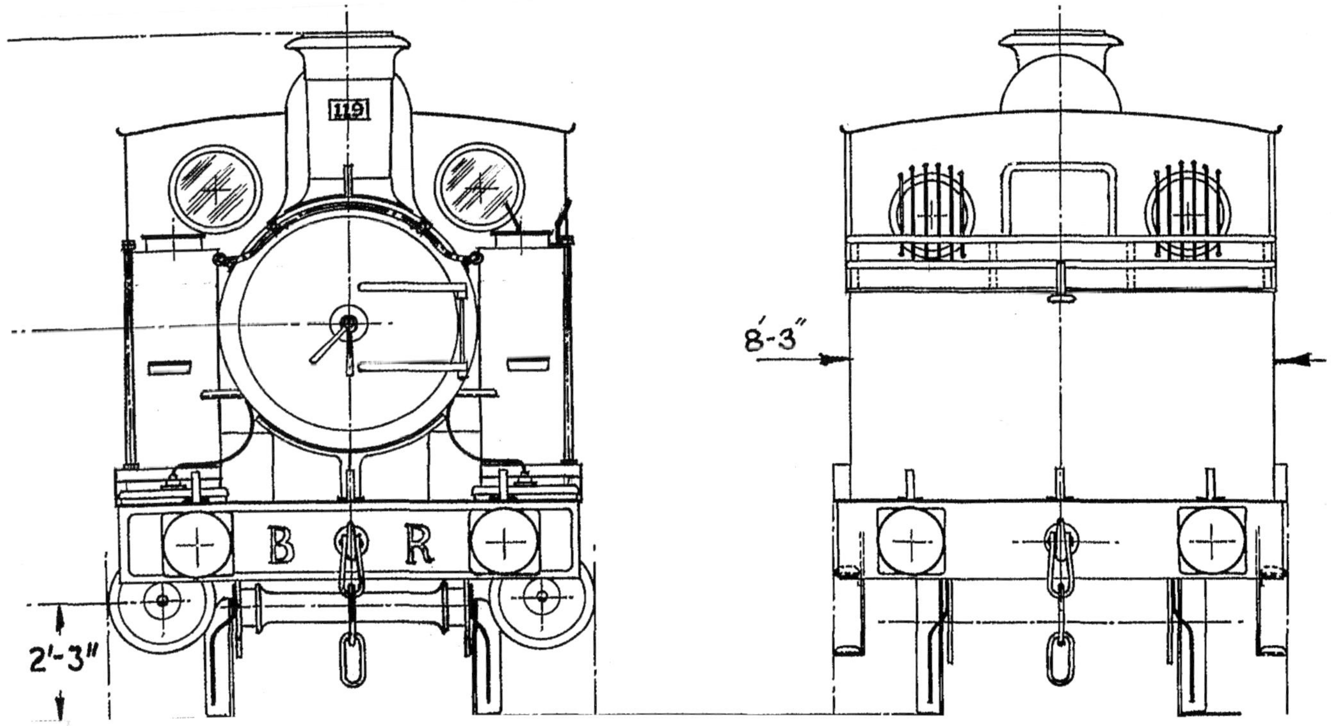

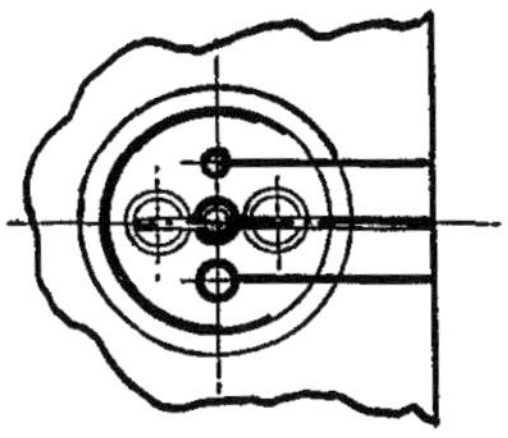

PLAN ON SAFETY VALVE SHOWING POSITION OF TWO BELL TYPE WHISTLES.

Class K Specification

Maker	Cooke Locomotive & Machine Co, Paterson, New Jersey
Numbers	117–121
Maker's numbers	2484, 2482, 2483, 2485, 2486
Date built	1899
Wheel arrangement	0-6-2T
Cylinders	18 in. × 26 in.
Wheels Coupled Trailing	 4 ft 3 in. 3 ft 6 in.
Wheelbase	7 ft 0 in. + 7 ft 5 in. + 6 ft 3 in.
Tractive effort	22,030 lb
Bunker capacity	35 cwt, 45 cwt after rebuilding by the GWR
Tank capacity	1,600 gallons
Weight full	56 tons 5 cwt

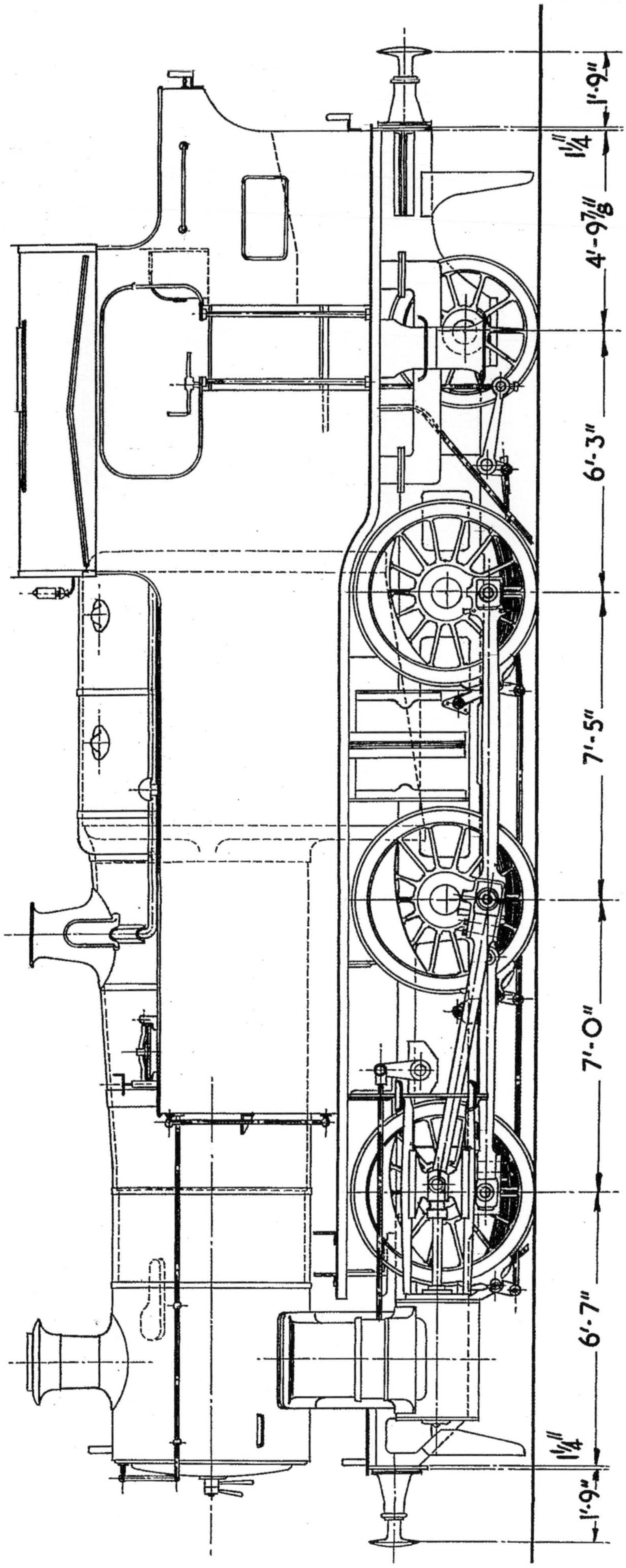

Drawing BRL18: 0-6-2T Class K Locomotive as rebuilt by the GWR

Only Nos. 118 and 120 (GWR Nos. 194 and 196) were rebuilt.

This locomotive, intended to be a substitute for the Class B1, had been a disappointment to the Barry Railway. Its costly performance in hauling mineral traffic, in terms of coal and water consumed, meant that it ended its days on the Barry on banking duties with one used as shed pilot at Barry.

It was somewhat of a surprise, therefore, when two of the class, No. 118 (GWR No. 194) and No. 120 (GWR No. 196) were sent to Swindon in 1923 for modifications. They were completely rebuilt with GWR Standard No. 3 superheated taper boilers, new tanks, high-roofed GW cabs and rebuilt bunkers. They returned to traffic duties in 1924. The class was given a GWR classification of Red B for the unmodified locomotives and a Red C for the modified ones.

They all returned to Barry in 1925 being confined once again to banking duties and short freight trips. In a five year period from 1927 to 1932, the class was withdrawn, condemned and cut up in Swindon.

Renumbering

Barry Railway numbers	Great Western Railway numbers
117–121	193–197

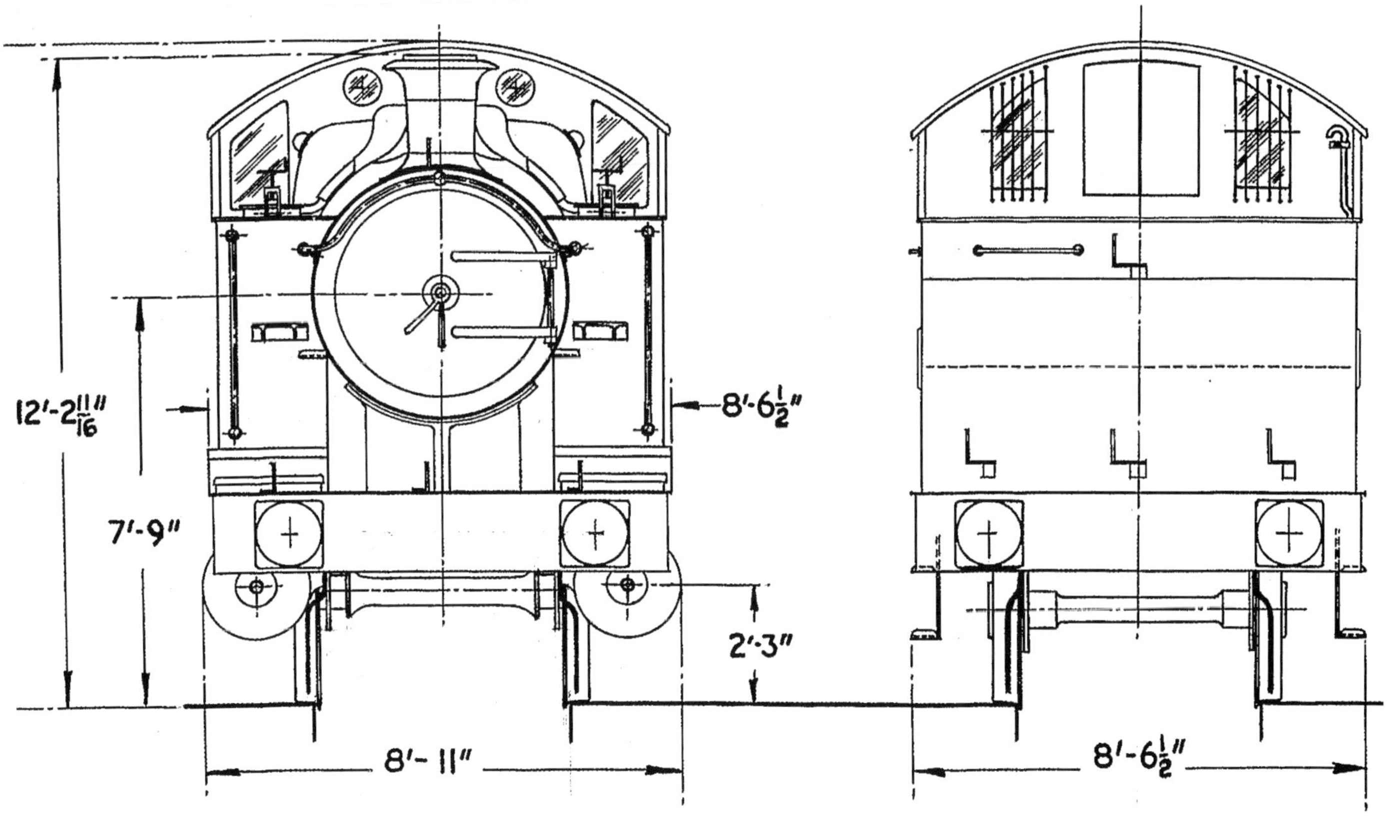

Class K No. 119. Although the Class K was as powerful as the B1, its uneconomical use of coal and water meant that it ended its days on non-mainline duties, one of which was hauling coal trains from Cadoxton Yard to the coal tip roads above Barry Docks (see the photograph on pages 76 and 77).

LOCOMOTIVE PUBLISHING CO/REAL PHOTOGRAPHS 15853

Class K No. 119. Given that these locomotives were manufactured in the United States, it is not surprising that they were known as "Yankees" among the enginemen. They were actually sent to Wales in kit form and assembled in Barry Locomotive Works located within the docks not far from the goods depot. No. 118 was the first kit to arrive in October 1899. PHOTOGRAPHER UNKNOWN

Class K No. 117. The Class K was the third Barry Railway locomotive to have outside cylinders, following the Class D and the Class H. Outside cylinders were regarded as being a standard of American locomotive design. KEN NUNN COLLECTION H1019/LCGB 26 APRIL 1910

Barry Railway Locomotive Works – the stone tablets on the gable ends of the roof show "B.R." on the left and "1887" on the right, thus dating the construction of the building. No new locomotive was ever constructed at the works, though the Class K which arrived in kit form from Paterson, New Jersey in 1899 was assembled here. In addition, in the previous year of 1898, the Works undertook to convert the Class C from its original wheel arrangement of 2-4-0T to its new 2-4-2T arrangement. The line to the right, on which the wagons are standing, led to a traverser at the rear of the Works by which locomotives could be moved to different locations within the Works complex. LOCOMOTIVE & GENERAL RAILWAY PHOTOGRAPHS RES 23791

Barry Railway

Committee Minutes

Committee of the Whole Board
1 March 1899

"Tenders for 10 and 20 additional locomotives were submitted and it was resolved to recommend that the tender of Messrs Sharp Stewart & Co for 10 engines, at the price of £2,800 be accepted.
Mr Hosgood was instructed to visit certain makers in Belgium and it was resolved to recommend that tenders be obtained from American and Belgian firms for 10 additional engines."

Committee of the Whole Board
5 April 1899

"Tenders for 10 additional locomotives were submitted and it was resolved to recommend that the following be accepted:-

Cooke & Co:	5 locomotives at £2,600 each;
Societe Anonyme Franco-Belge:	5 locomotives at £2,700 each.

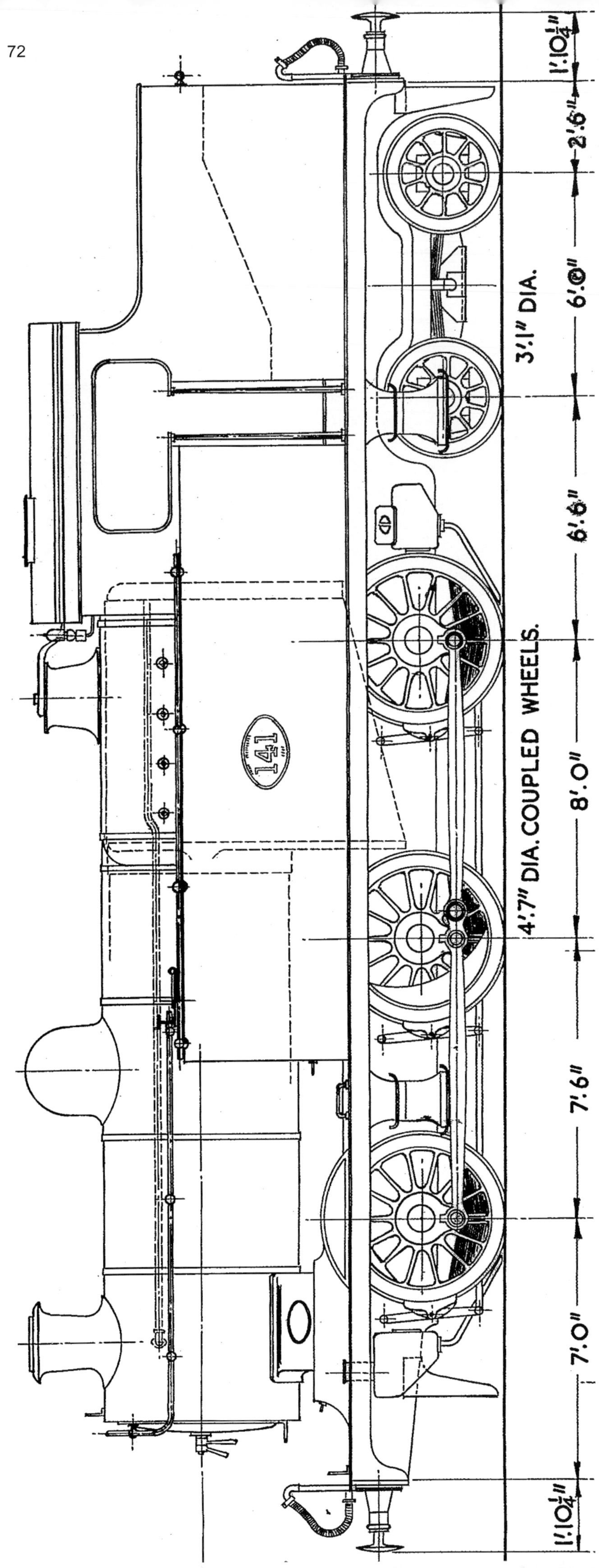

Drawing BRL19: 0-6-4T Class L Locomotive

This was Mr Auld's only design for the Barry Railway and the company's last locomotive of which 10 were ordered in 1914 from Hawthorn, Leslie. In many ways, it was a radical departure from traditional Barry Railway practice. Auld introduced new standards for the boiler, cylinders and wheels. It was a much larger, Belpaire type, boiler which necessitated a shorter chimney. On all other classes of Barry locomotive, the engine number was to be found on the front of the chimney. The shortness of the Class L chimney meant this was not possible and so was located on the smokebox door.

It had however a design fault which rendered shunting with the Class L subject to serious impediment. The rear coupled wheels forced open the tongue of hand-operated points. This resulted in the bogie wheels turning down the wrong track leading to derailment, and mainly occurred in the docks area or colliery sidings. This was only a problem when the engine was travelling chimney first. The most devastating accident occurred in 1921 when No. 147 derailed and turned on its side in Barry sidings while hauling a coal train.

Several solutions were attempted including allowing for more sideplay to the coupled axleboxes as well as lengthening the tongues of offending points. In practice, the fireman would have to hold the point lever while the locomotive inched its way over the point in question.

Although designed for coal traffic and known as "Dreadnoughts" by the enginemen, the class was also vacuum-fitted for passenger traffic and was occasionally used on passenger services to Cardiff. One publicity photograph taken in 1920 shows No. 148 at the head of the seven-coach suburban set. However this was not a common sight on the Barry.

The class's suitability for long-distance mineral traffic was exemplified by the fact that it was listed in the working timetables for hauling coal trains from Neath Junction and Rogerstone Yard and could also be found hauling mineral traffic from Rhymney and New Tredegar over B&M rails to Barry Docks.

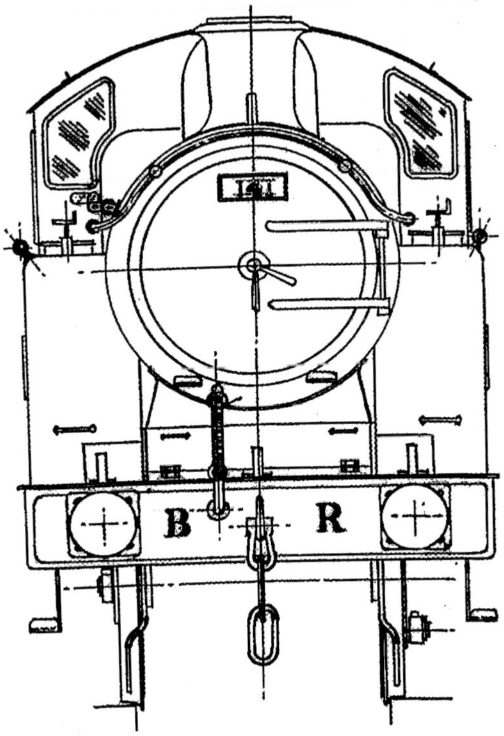

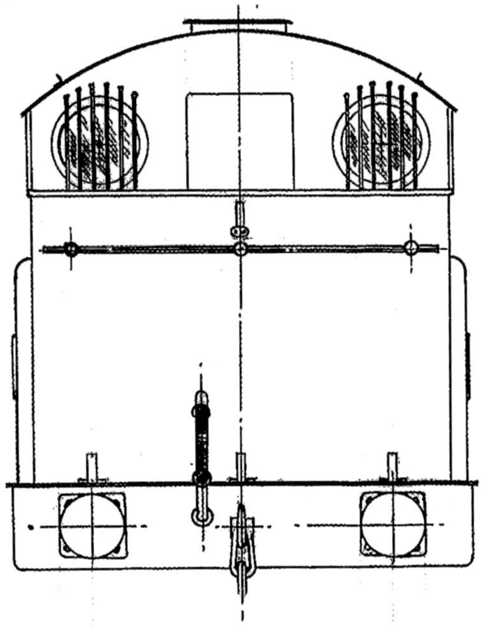

Class L Specification	
Maker	Hawthorn, Leslie
Numbers	139–148
Maker's numbers	3038–3047
Date built	1914
Wheel arrangement	0-6-4T
Cylinders	18½ in. × 26 in.
Wheels	
Coupled	4 ft 7 in.
Trailing	3 ft 1 in.
Wheelbase	7 ft 6 in. + 8 ft 0 in. + 6 ft 6 in. + 6 ft 0 in.
Tractive effort	24,775 lb
Bunker capacity	80 cwt
Tank capacity	2150 gallons
Weight full	74 tons 15 cwt

Class L No. 143. This was the first and last locomotive designed by Auld for the Barry Railway and the only one constructed by Hawthorn, Leslie. They were designed to be a mixed traffic engine and could be seen at the head of passenger services from Bridgend to Cardiff (see page 105). No original GA drawings of the locomotive have survived owing to a fire at the company's offices. HAWTHORNE, LESLIE

Class L No. 141. The wheel arrangement of 0-6-4T was unusual not only for the Barry Railway but also for South Wales in general; only the Alexandra Docks and Railway company had locomotives with the same arrangement. The additional double bogie was required to support the larger bunker which had a capacity of 4 tons. LOCOMOTIVE PUBLISHING CO./REAL PHOTOGRAPHS 15854

Class L No. 144 hauling a mineral train. This is one of only a few photographs showing a Barry Railway mineral train complete with coal wagons and brake van at the far end. The Class L introduced a novelty in having the locomotive number on the smokebox door rather than on the chimney or the front buffer beam. The Class L proved very effective at hauling mineral trains and could be found all over the system including the Brecon & Merthyr Extension as well as trains from Rogerstone and Neath Junction. H SHERVINGTON

The signal gantry in the top left-hand foreground frames Creigiau station. This gantry carried the Starter (No. 3 lever), Down Loop No. 1 (No. 7 lever) and Down loop No. 2 (No. 10 lever). A Class J waits to depart on the down main line; the train has only just arrived as the Home signal, located on the gantry to the rear of the station, is still off. The loops were used to park mineral or goods trains to allow passenger trains to pass through. In the case of Creigiau the loops were to the south of the station with the up side having a refuge siding rather than a loop. FROM A CONTEMPORARY POSTCARD

Class J No. 86 with Class K and Class D in the background. This scene brims with railway activity and represents the breadth of activity that was the Barry Railway at its zenith. Centre stage is Class J No. 86 hauling a 9-coach rake on the Barry to Cardiff commuter route not far from Biglis Junction. Behind in Cadoxton Yard is an unidentified Class D which may either have arrived with a train from the B&M extension or is preparing to take a train to Barry Docks sidings. To the left is a Class K which more than likely is on shunting duties between the Yard and the coal tip sidings. HEBER SHIRVINGTON, STEPHEN ROWSON COLLECTION

INSOLES

Drawing BRM1: Motor Cars

The Barry Railway was one of the first to consider the use of steam Motor Cars, primarily for use on the Cardiff to Pontypridd line. On 4th February 1904, the relevant committee minutes noted that:

"The Locomotive Superintendent was instructed to get out a tracing of a motor car suitable for this Company's traffic and to submit it at the next meeting."

Hosgood, the Locomotive Superintendent from 1888 to 1905, wasted no time and submitted a detailed report on the 29th February. In it, he described his visits to the Wolseley Motor Car Company at Birmingham and to the North Eastern Railway Company's works at York. The purpose of the first visit was to inspect an 80hp petrol engine used to power the railcar and, for the second visit, to see the railcar at work. The railcar had been the subject of experimental running for the previous 12 months while various engines had been tried. The Wolseley 80hp engine was the third to be tested after other previous failures.

The total weight of the car was 35 tons and there was no luggage accommodation or divisions for classes. A large portion of the length was taken up, in Hosgood's view, by extremely complicated and delicate machinery. The claims made for the railcar were that it had a maximum speed of 30 mph and that it could climb a 1 in 80 gradient at a speed of 20 mph. On this last claim, Hosgood was sceptical.

Hosgood concluded that a petrol driven railcar was not suitable for the Barry Railway because of its unreliability. The sensitivity of the machinery meant that the engine could fail if it was not carefully operated and that the person driving would have to be a highly trained specialist rather than "an ordinary artisan". His preference was rather for the steam rail motor being developed by the Great Western Railway which had been running successfully between Stonehouse and Chalford in Gloucestershire. The Barry required a means of propulsion that could replace locomotive-hauled trains. In his view, however, it was still at an experimental stage with improvements continually being made and therefore it might suit the Company to defer the matter for a short while.

On this last point, the Board, known for its pioneering spirit, disagreed. On the 3rd March 1904, it was resolved that the Locomotive Superintendent, John Hosgood, and Manager, Richard Evans, provide designs for a suitable car and submit them to the next meeting. By June of the same year, tenders had been submitted and the Board accepted the submission of the North British Locomotive Co for two Motor Cars at a price of £2,400 each, with an option for a further two, if required, at the same price. NBL built the power units but the carriage units were constructed by R Y Pickering & Co of Wishaw.

Just under a year later, on 1st May 1905, the Motor Cars had been received and had commenced operations on the Cardiff to Pontypridd service that morning. It was not long however before problems began to emerge and the Company started to receive complaints from passengers. As Hosgood stated in his report of 29th May 1905:

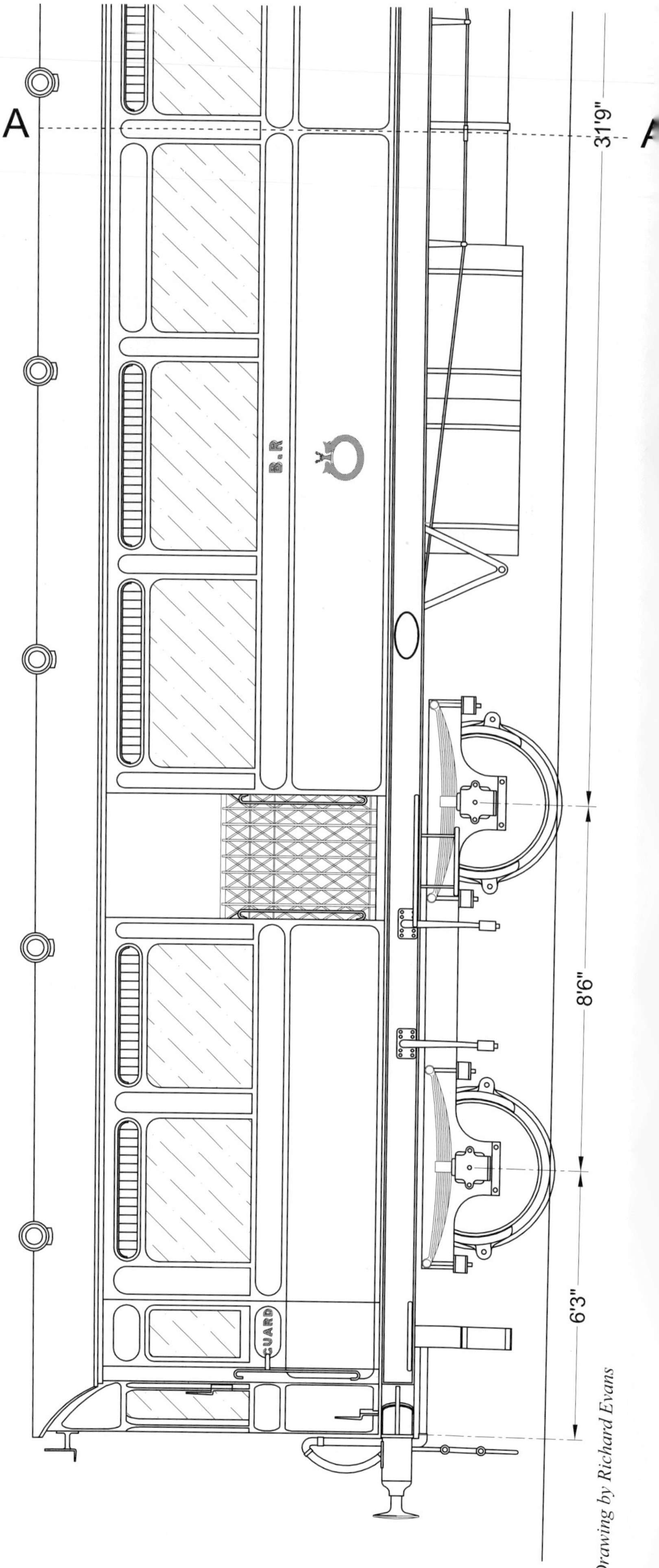

Drawing by Richard Evans

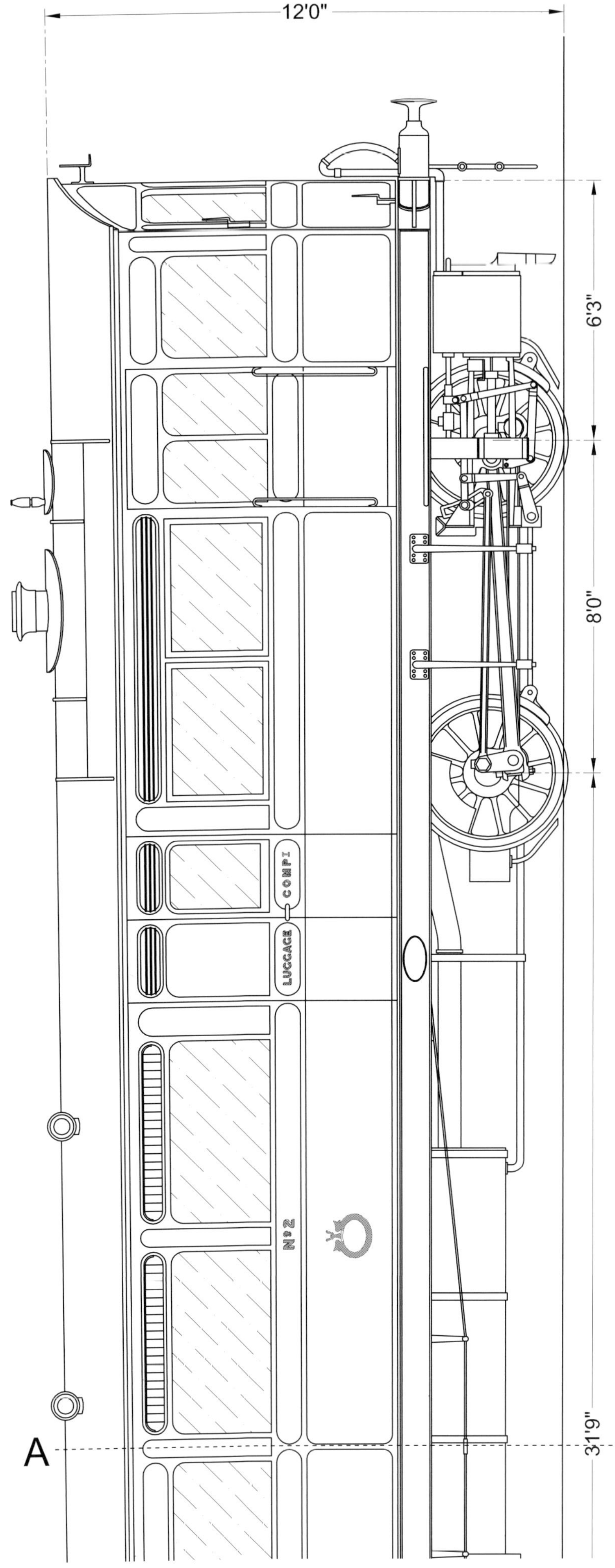

"After testing the Motor Cars for a fortnight I am satisfied that they are not able to do the Cardiff and Pontypridd service and give public satisfaction. I have received numerous complaints from passengers travelling between Cardiff and stations on the Main Line, as well as two petitions asking that the train service be reinstated.

"Summarised shortly, the complaints of passengers refer to:-

1. The alteration of the times
2. The reduced speed
3. Late running
4. Insufficient room at certain times."

One side effect of this was the slowing up of delivery of parcels at Pontypridd, a business that was growing in importance for the Barry, with customers threatening to take their business elsewhere if the Motor Car service continued. It was clear to Hosgood that replacing a fast locomotive-hauled passenger service with a slower Motor Car service had been rejected by the travelling public. He therefore took it upon himself to reinstate the old service from 1st June 1905; the steam Motor Car service had lasted no more than a month.

They did not however become redundant. As Hosgood continued in his report:

"... I have arranged to supplement the train service on the Vale of Glamorgan Railway with a motor service between Barry and Llantwit and Southerndown Road, these being the points between which the heaviest traffic is passing."

Thus the Vale of Glamorgan became their final place of operation until their conversion to semi-corridor trailers in 1914 (see page 82). Renumbered 177 and 178, they were coupled together to become the Vestibule Set which operated between Barry and Bridgend, hauled by a Class C. They were finally withdrawn and condemned in 1951.

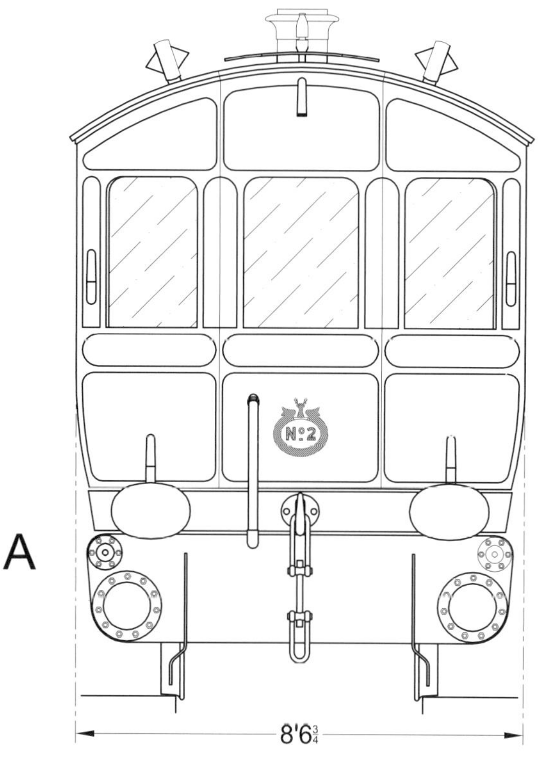

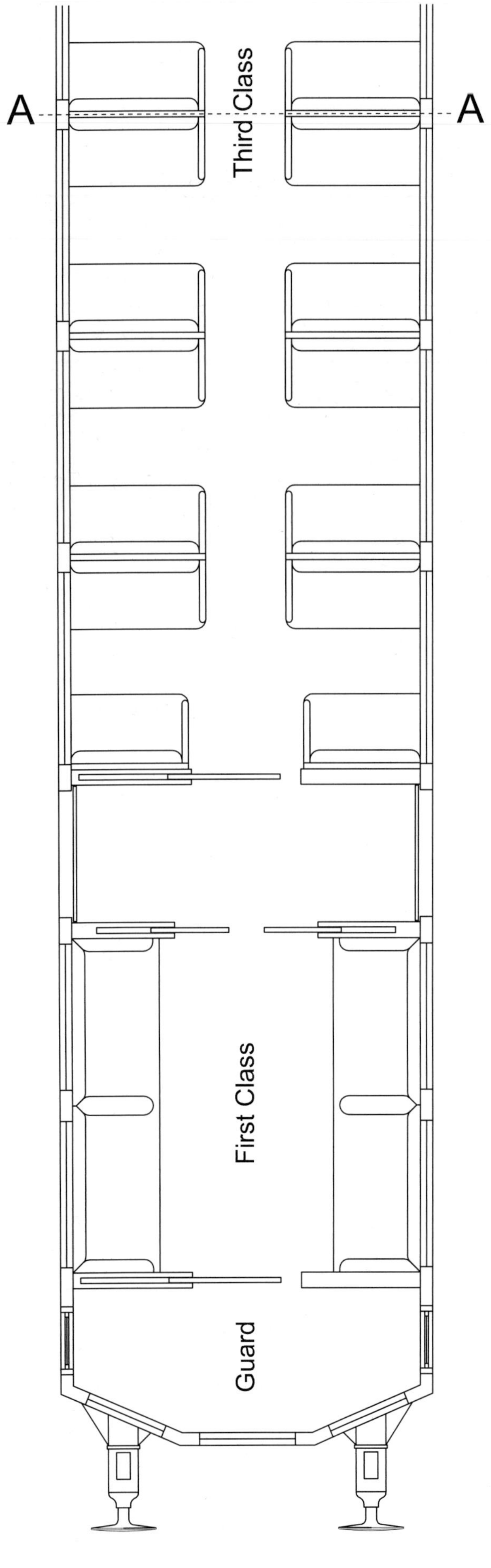
A
A
Third Class
First Class
Guard

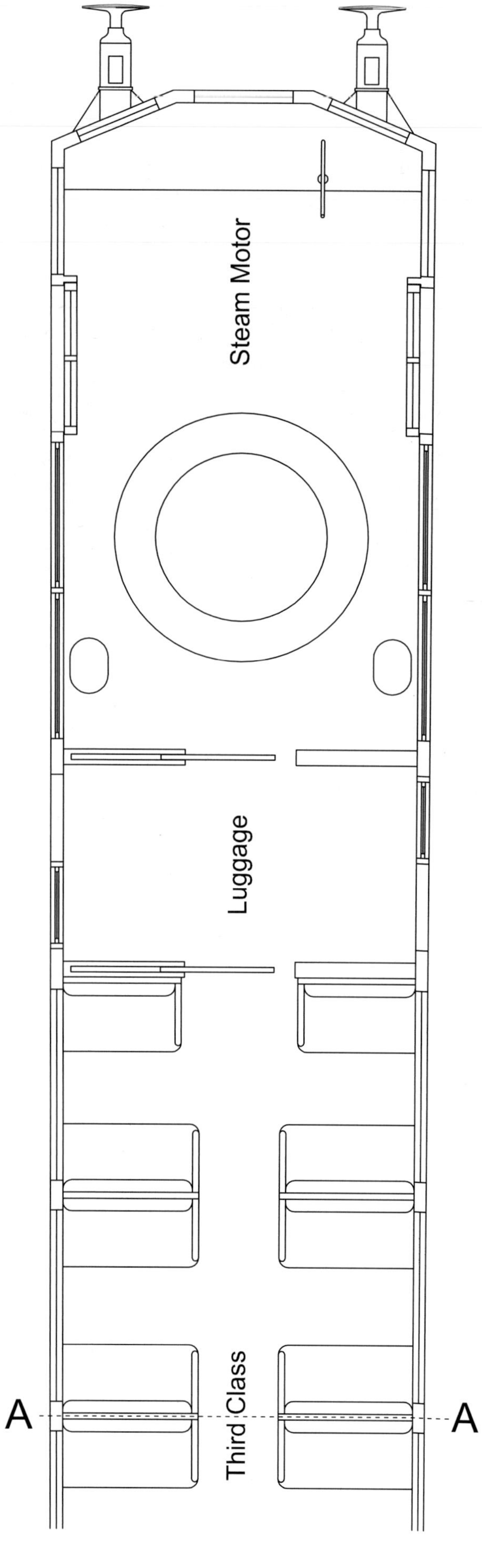
Steam Motor
Luggage
Third Class
A
A

The Barry Railway had two motor cars (outside of South Wales, they were known as steam railmotors) of which No. 2 is shown in the picture. They were delivered in 1905 and intended for service on the Cardiff to Pontypridd route. The doorway, which had retractable steps, was to the rear of the car away from the power unit. The First Class compartment with ten seats was located to the rear of the doorway area just ahead the trailing driving compartment. The Third Class compartment with 40 seats was forward of the doorway area and behind the luggage compartment. Ahead of this was the motor compartment including the leading drivers' compartment. R Y PICKERING

Above: *This is the end view of Motor Car No. 2 showing a number of interesting features. There are lamp brackets at the centre top of the Car as well as two others located above the buffers. On Motor Car No. 1, the lower centre panel carried the Barry Railway armorial device. No. 2 does the same except that the Welsh dragon has been replaced by the number of the Car.* R Y PICKERING

This is No. 2, pictured just outside of Barry Carriage Shed; the sharp curve to Barry Island can be seen to the left. Here the Car is sporting a lamp on the top centre bracket and looks resplendent in its lake livery with gold lining. The water capacity of the Cars was 500 gallons and each unit carried 15 cwt of coal. J PITT COLLECTION

The drawing shows Motor Car No. 2. As well as the detail differences in the livery of the two motor cars, there were differences on the roof of the power end. Motor Car No. 2, as shown in the drawing, had circular protective plates around the chimney and whistle which protected the roof beneath. No. 1, however, had plates which were square with rounded corners. The reason for the variation is unknown, but may reflect experiments to address operational issues once the cars entered traffic.

BARRY RAILWAY MOTOR CAR SPECIFICATION

Length (over headstocks):	60 ft 10½ in.
Width:	8 ft 11 in.
Height (rail to roof):	12 ft 0 in.
Wheelbase:	48 ft 3 in..
Power Bogie Wheelbase:	8 ft 0 in.
Wheel diameter	3 ft 7½ in.
Boiler length:	9 ft 1 in.
Boiler diameter:	4 ft 11 in.
Boiler pressure (maximum):	160 lb/in^2
Cylinders:	12 in. diameter × 16 in. stroke
Tractive effort at 90%:	7,680 lb
Water capacity:	500 gallons
Coal capacity:	15 cwt
Total working weight:	50 tons 15 cwt

Drawing BRM2: Motor Cars after Conversion to Hauled Stock

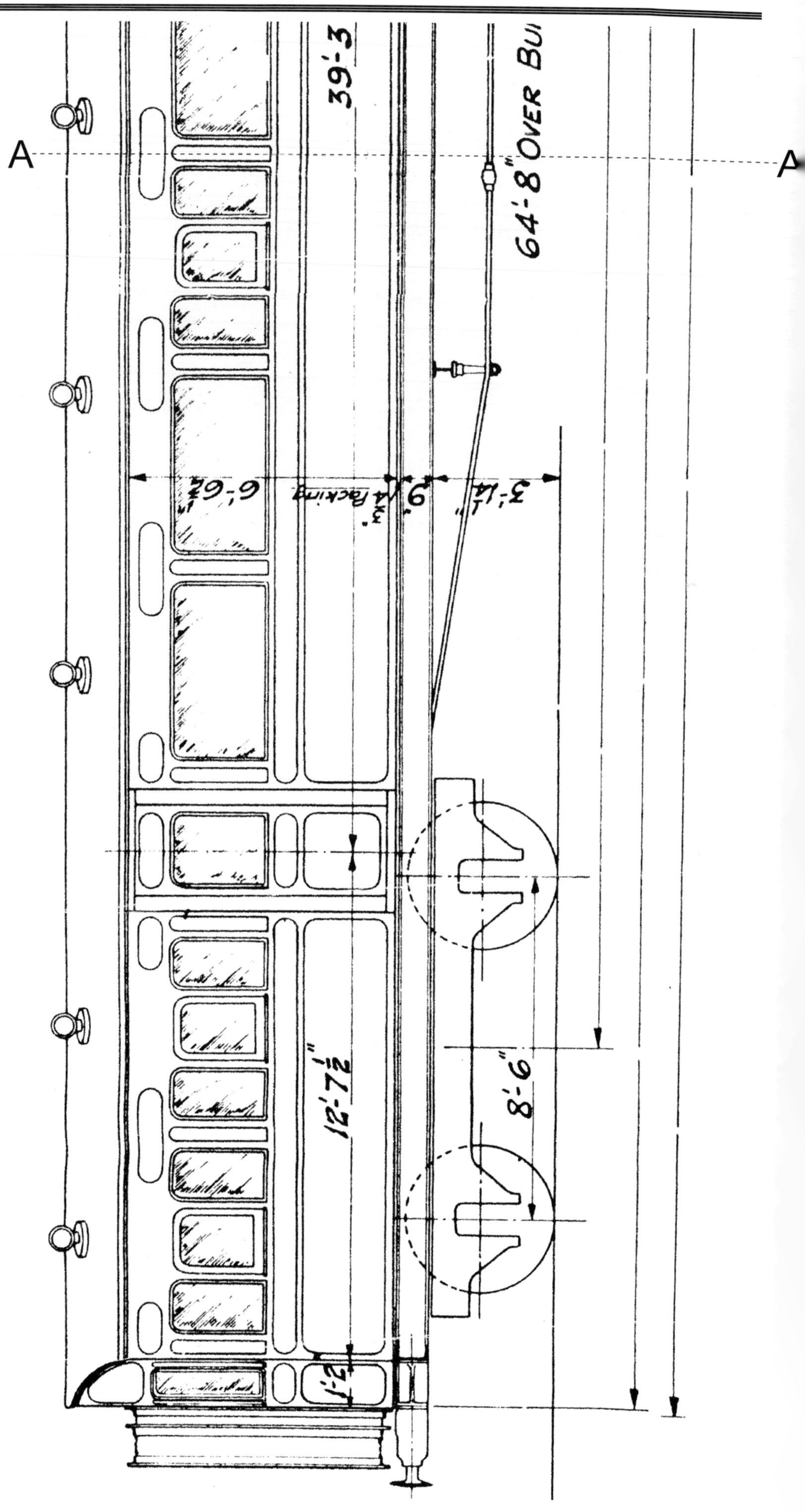

Drawing BRC1
Third Class Carriages Nos 27–41

Of the six-wheeled carriages, by far the largest number, 75 in all, were Third Class. A number of manufacturers were used to produce these: Brown, Marshall, Gloucester Railway Carriage & Wagon Co, Ashbury and Oldbury as well as second-hand stock purchased from the Lancashire & Yorkshire Railway in 1916, the only L&Y stock to be sold to another company.

The particular example in the drawing was built by the Gloucester Railway Carriage & Wagon Co in 1895 and was 33 ft 11 in. in length with six Third Class compartments. They were to be found throughout the Barry Railway network.

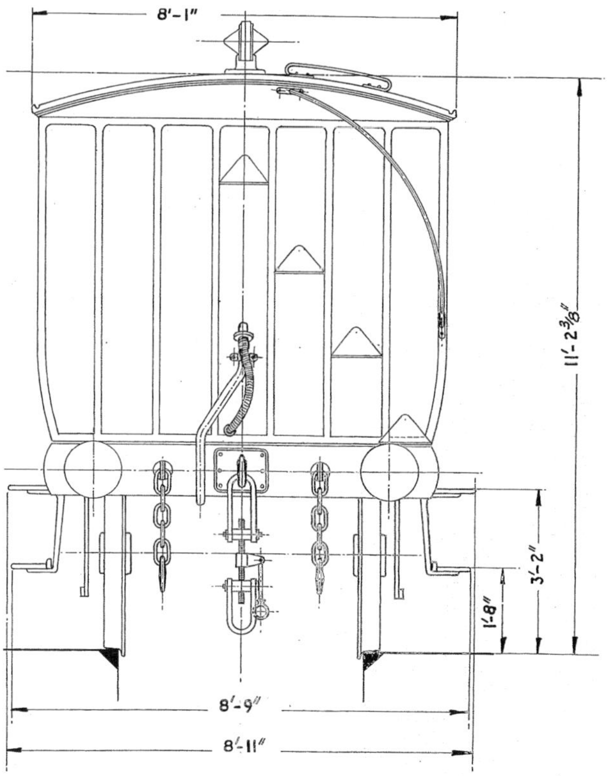

A three quarters view of a six-wheeled Third, No. 27. It was built by the Gloucester Railway Carriage and Wagon Company in 1895 and had six Third Class compartments, being 33 ft 11 in. in length. In total, Gloucester built 15 of these carriages for the Barry Railway, though two, Nos. 37 and 38, were later converted to Brake Thirds in 1909. The Brake Third was close coupled to a four-wheeled Composite coach to form a "Motor Set" of which the Barry Railway possessed two. Coaches No. 152 and No. 157 formed the Composite halves of the Sets.
Gloucester Railway Carriage & Wagon Co, Gloucestershire Archives

Third No. 36 came from the same batch as No. 27 above. The six Third Class ompartments can be seen more clearly with two Barry Railway armorial devices at either end of the carriage. By comparing these two photographs, it is possible to see that the access steps to the roof were provided at one end of the carriage only.
Gloucester Railway Carriage & Wagon Co, Gloucestershire Archives

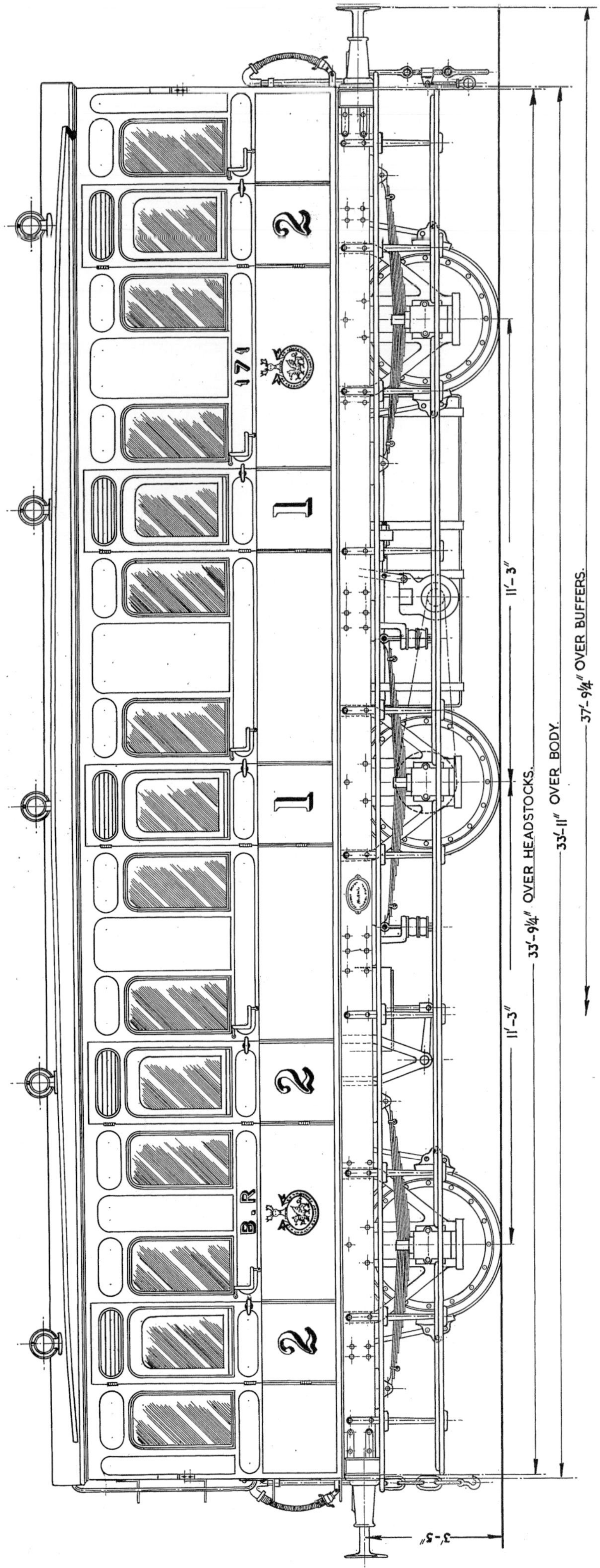

Drawing BRC2
First/Second Composite Carriages
Nos. 158–174

The vast majority of the Barry Railway's coaching stock comprised six-wheeled carriages, 141 in total. The example in the drawing was a Composite, 33 ft 11 in. in length, built by Ashbury between 1895 and 1898; an additional two, Nos. 175 and 176, were built by Oldbury in 1900. They comprised two First Class compartments and three Second Class compartments. There were two types of steps for roof access. They were employed on the carriage end for this particular Composite. Some carriages had steps from both sides of the coach while others had steps from one side only.

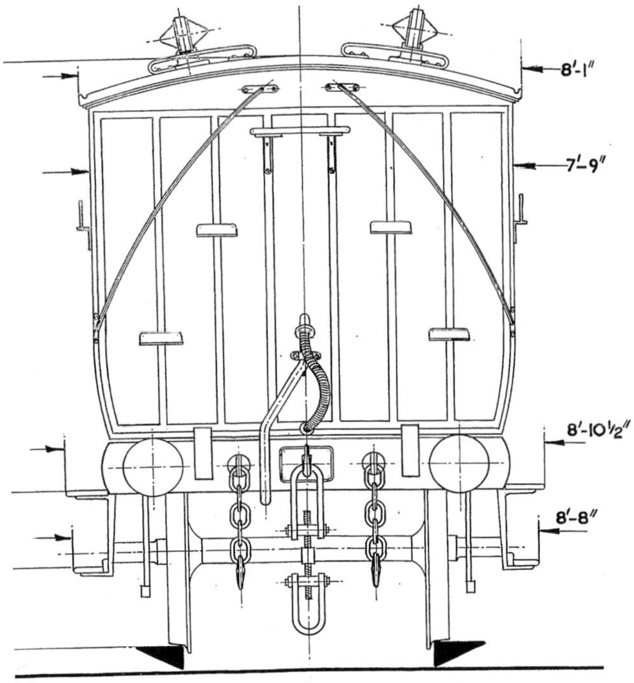

Minutes of the Capital Expenditure Committee

7 August 1894

The Secretary submitted a summary of tenders (for passenger carriages) received which had been opened by Mr Hood.
It was left to Mr Davies and the Manager to make enquiries and hereafter to accept one of the two lowest tenders, *viz* the Ashbury Carriage & Wagon Co and the Gloucester Railway Carriage & Wagon Co.

The ubiquitous passenger Class J is seen here at the head of a four-coach rake at Creigiau station, waiting to head south on a Pontypridd to Cardiff Clarence Road service. It was somewhat unusual but not unknown for a Class J to be seen on main line passenger duty. Both the Barry to Porth and Cardiff to Pontypridd services were generally hauled by Class Gs. Unfortunately, no date is given for this photograph but it is after 1899 since that was the year when No. 97 was delivered. Interestingly, sometime between 1905 and 1909, this locomotive lost its brass chimney cap which was replaced with a copper version.
PHOTOGRAPHER UNKNOWN

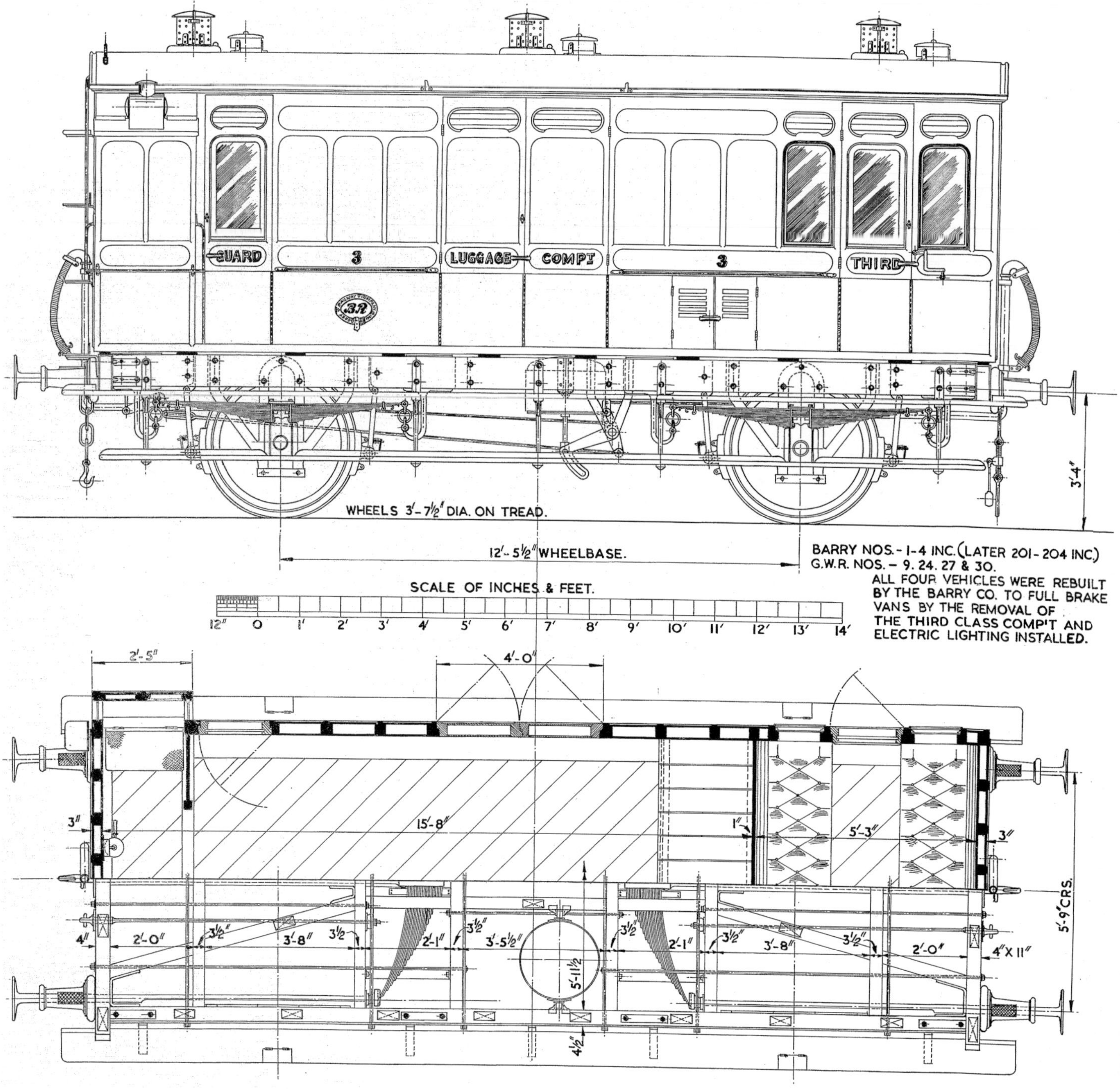

Drawing BRC3 Brake/Third Carriages Nos 197–204 Originally numbered 1–8 (but see text)

The Barry Railway began its passenger services with 4-wheeled carriages which were built in 1888 by the Metropolitan Carriage & Wagon Company. They appear to have been numbered according to the different categories of coach. The Barry started with nine Thirds, eight 1st/2nd Composites and four Passenger Brake Vans. So there were Thirds 1–9, Composites 1–8 and Brakes 1–4. These were subsequently renumbered using a sequential system so Thirds 1-9 remained as Nos. 1–9, Composites 1–8 became Nos. 150–157 and Brakes 1–4, according to Mountford, were renumbered 197–200. This is, however, at odds with the notes on the drawing itself which state that 1–4 were renumbered 201–204. There is another anomaly. According to the drawing, the coach was built in 1889. However, Mountford shows numbers 197–200 as having been built in 1888 and Nos. 201–204 were built in 1890. The third difficulty is that numbers 197–204 were built as Brake Vans. Only one of these was converted by the Barry by having a Third Class compartment installed by shortening the Luggage compartment. This was No. 197 which would have originally been No. 1, according to Mountford. Therefore it would not be possible for the carriage to be No. 3 in the series as is shown on the drawing.

To further complicate matters, Chapman made notes on the numbering and renumbering of the Barry coaches and these support the notes shown on T L Jones' drawing. Further research is required to resolve this issue.

As for the coach itself, it was 21 ft 6 in. in length, as were all the Passenger Brake Vans manufactured by the Metropolitan Carriage & Wagon Co, with one Third Class compartment 5 ft 3 in. in width and one Luggage compartment which was separated from the guard's compartment by a sliding door. This and the other four-wheelers would have been used on the Cogan branch between Barry Dock and Cogan from the opening of the line on 20th December 1888, later extended to Barry on 8th February 1889.

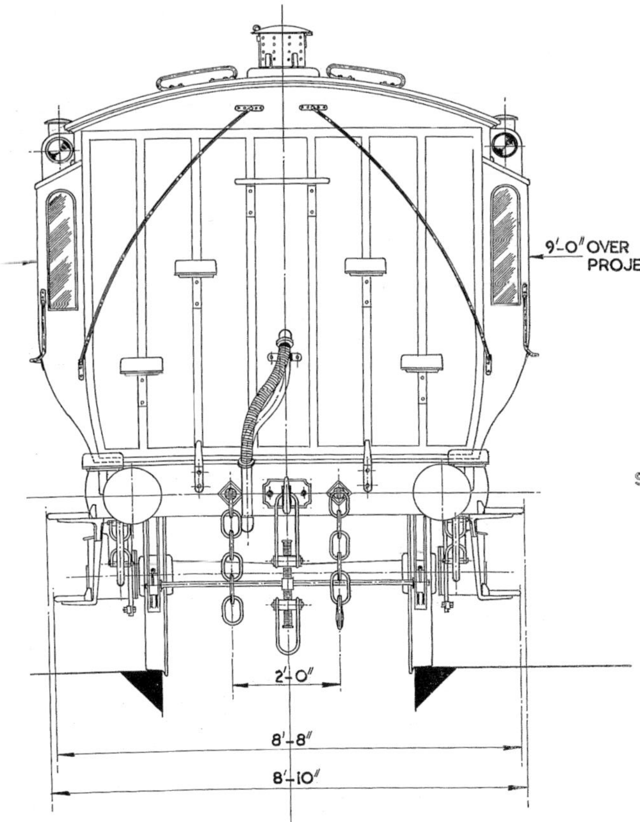

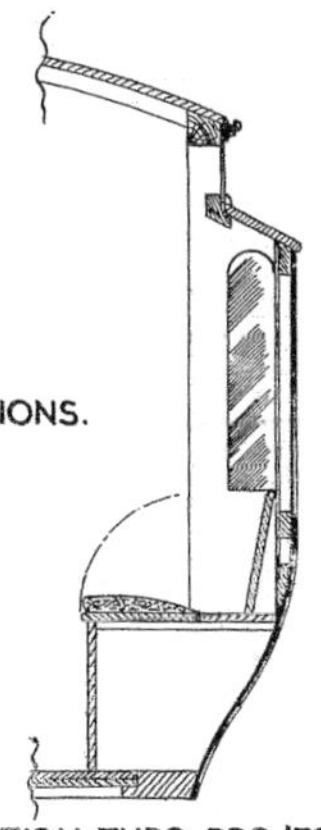

SECTION THRO PROJECTION.

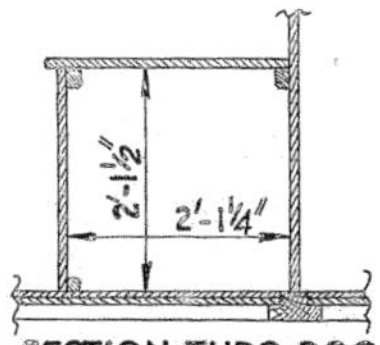

SECTION THRO DOG BOX.

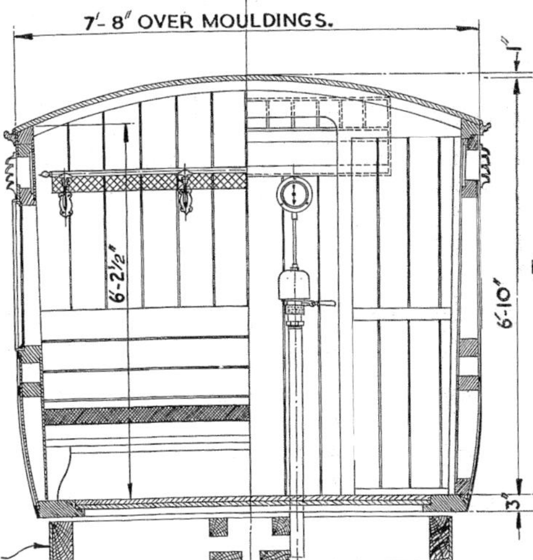

TIMBER SOLE FACED WITH 9"X 4" R.S.ANGLE.

SECT'N THRO' COMP'T. SECT'N THRO LUGG. COMP'T.

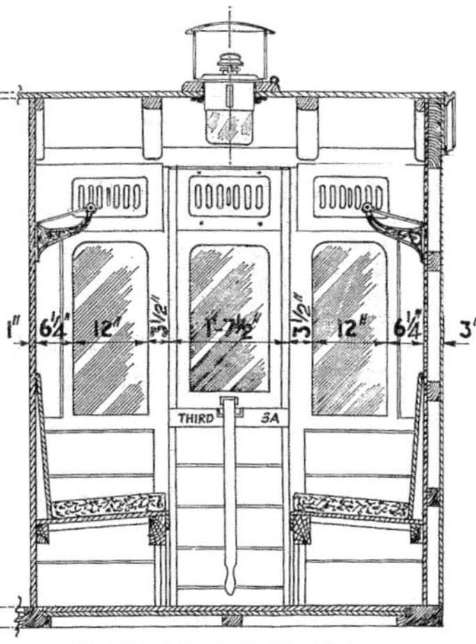

SECTION THRO' COMP'T.

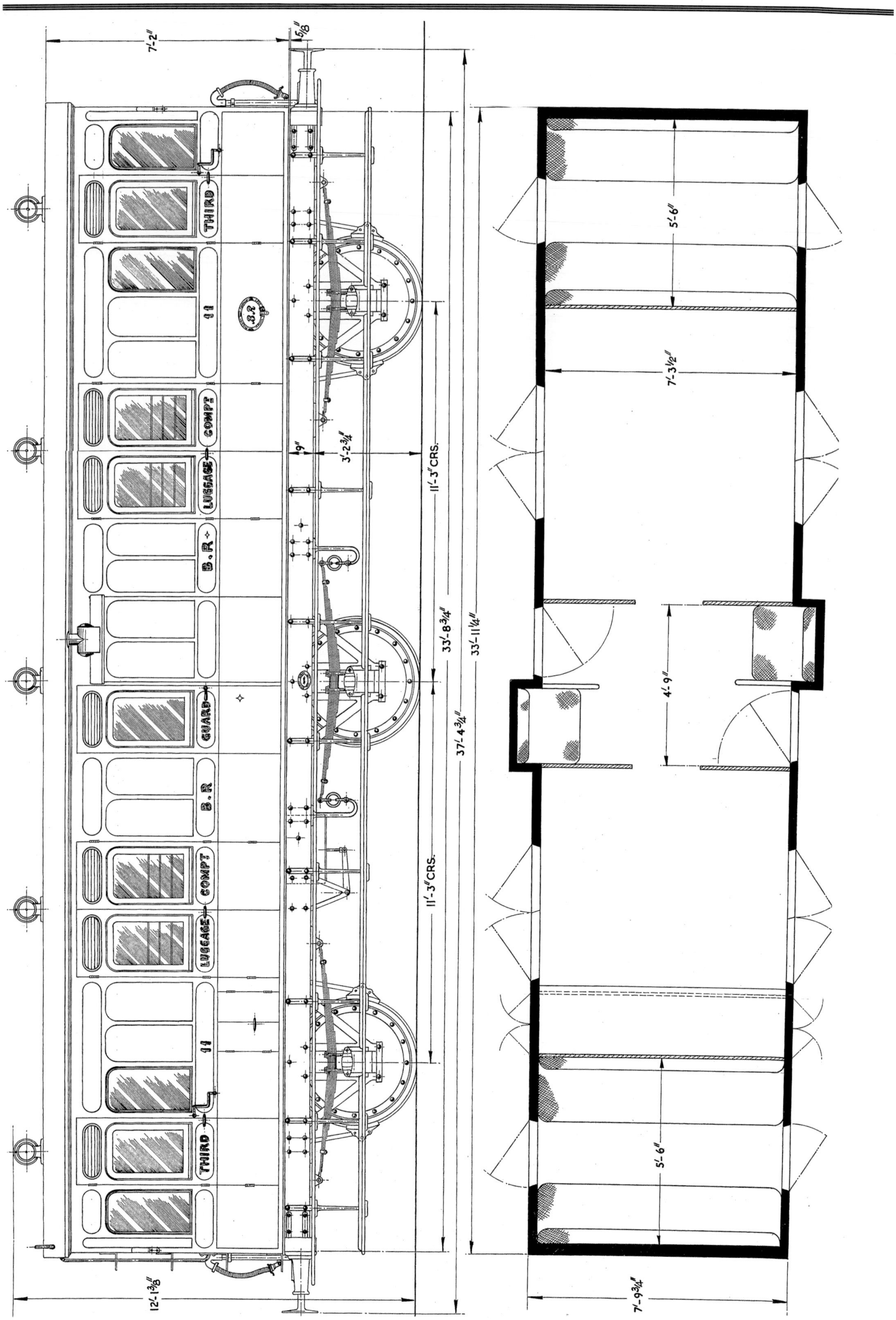
7'-2"
5 1/8"
THIRD
LUGGAGE
COMPT
B.R.
GUARD
9"
3'-2 3/4"
11'-3" CRS.
33'-8 3/4"
37'-4 3/4"
33'-11 1/4"
12'-1 3/8"
5'-6"
7'-3 1/2"
4'-9"
7'-9 3/4"

DRAWING BRC4 BRAKE/THIRD CARRIAGES NOS 205–208
ORIGINALLY NUMBERED 9–12

Both the Gloucester Railway Carriage & Wagon Co and Ashbury were involved in manufacturing this particular type of Brake Third between 1896 and 1898. The drawing is of the Gloucester vehicles as originally built and the Ashbury carriages, numbered 209–220, were very similar. Subsequently the Barry lengthened one of the luggage compartments of this type leaving one Third Class compartment only. Four of the Ashbury built coaches, Nos. 216–217 and 219–220, were converted to full Passenger Brake Vans with two extended luggage compartments.

The coach illustrated in the drawing was eventually converted to a one Third Class compartment carriage. It was built by the Gloucester Railway Carriage & Wagon Co in 1895 and was 33 ft 11¼ in. long.

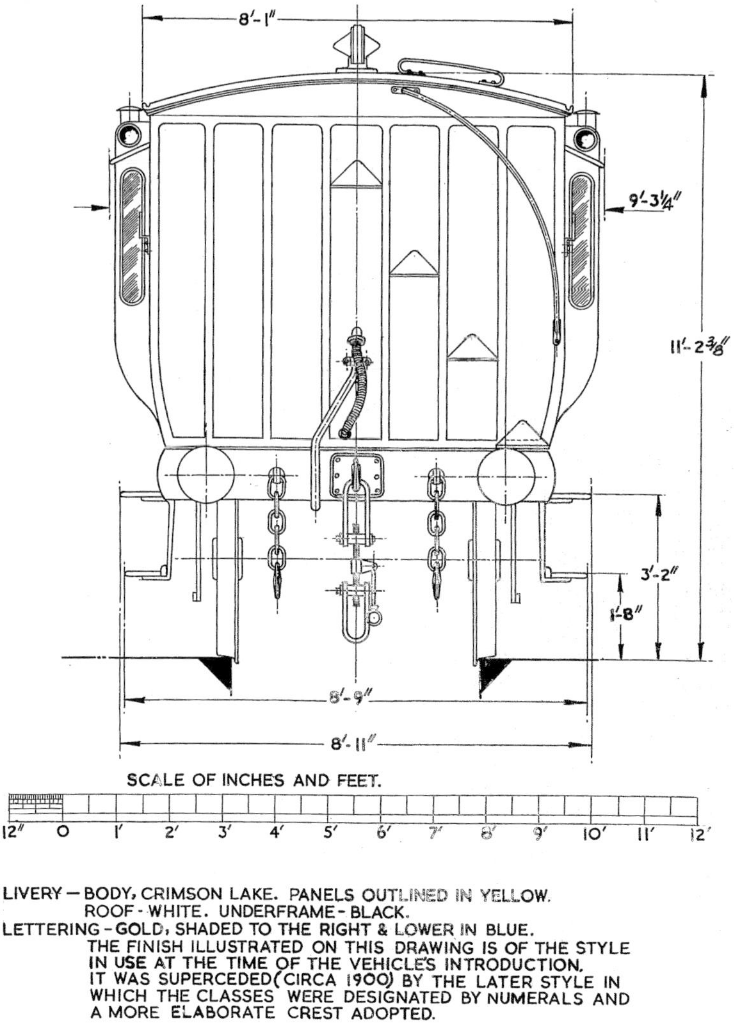

Between 1895 and 1900, 11 of this type, 3rd + Luggage + Guard + Luggage + 3rd, were built by various companies, the others being Ashbury and Oldbury. Gloucester built three of them in 1895. The original Barry Railway coach numbering system was based on the type of coach. For example, there would be Third 1 to 10, Composite 1 to 10, Brake Third 1 to 10 and so on. Around 1909, the Barry Company changed its system to sequential numbers, presumably to avoid confusion. Brake Third No. 12 has been numbered according to the old system. It was renumbered 208 under the new system. GLOUCESTER RAILWAY CARRIAGE & WAGON CO., GLOUCESTERSHIRE ARCHIVES

Drawing BRC5 Third Class Carriages Nos 2, 3, 4, 6, 8 and 10 (new series)

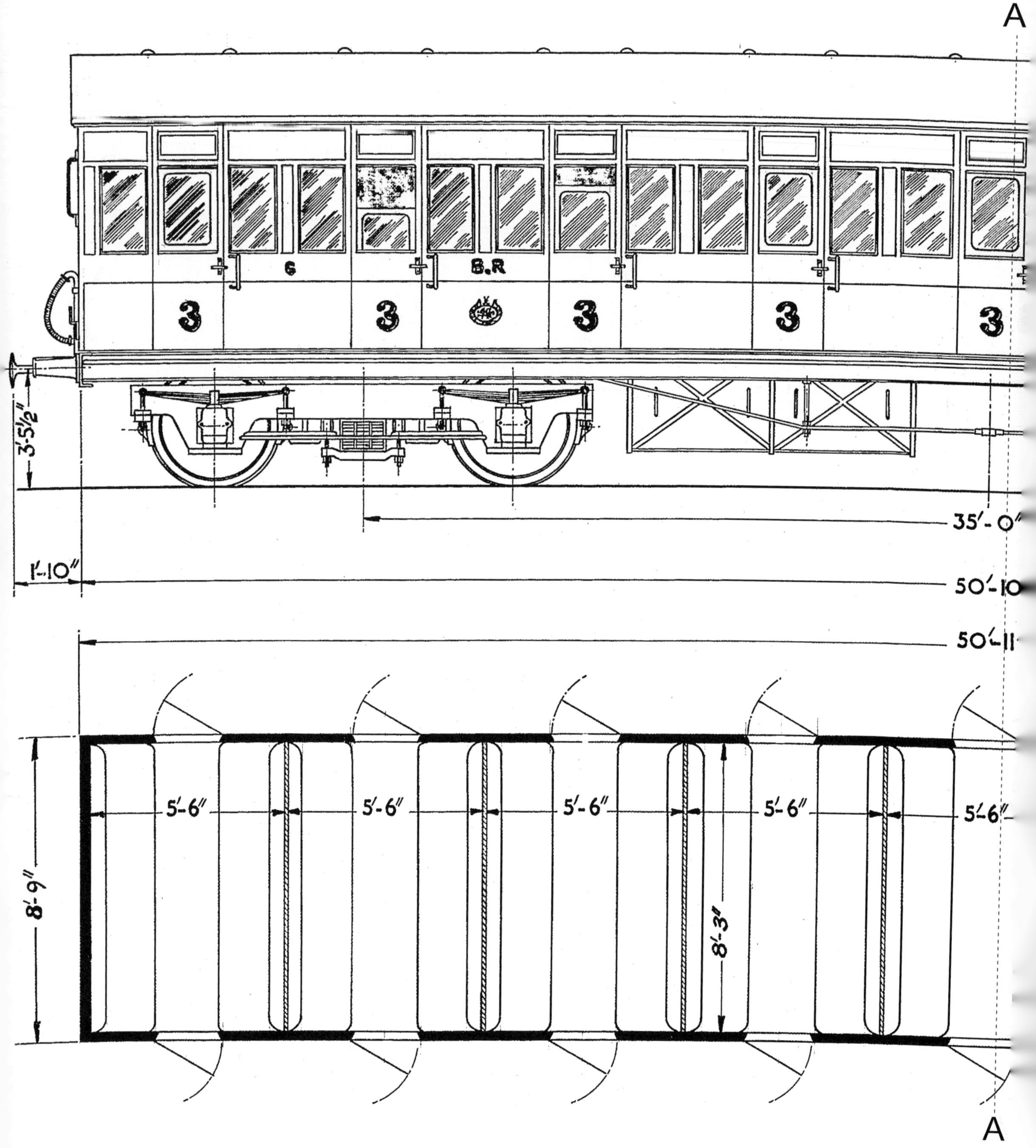

In 1920, as an apparent act of faith in the future, the Barry Railway ordered two seven-carriage sets of bogie carriages for use on commuter services to Cardiff. Each set comprised two Brake Thirds, two Tri-Composites and three Thirds. They were built by the Birmingham Railway Carriage & Wagon Company and were referred to informally as the "Marble Arch" set on account of the high-rounded nature of the roofs. All but two of the carriages survived into the 1950s, the last of which, Nos. 197 and 198, were withdrawn from service on 5th October 1962, having served out their days on the Culm Valley branch in Devon.

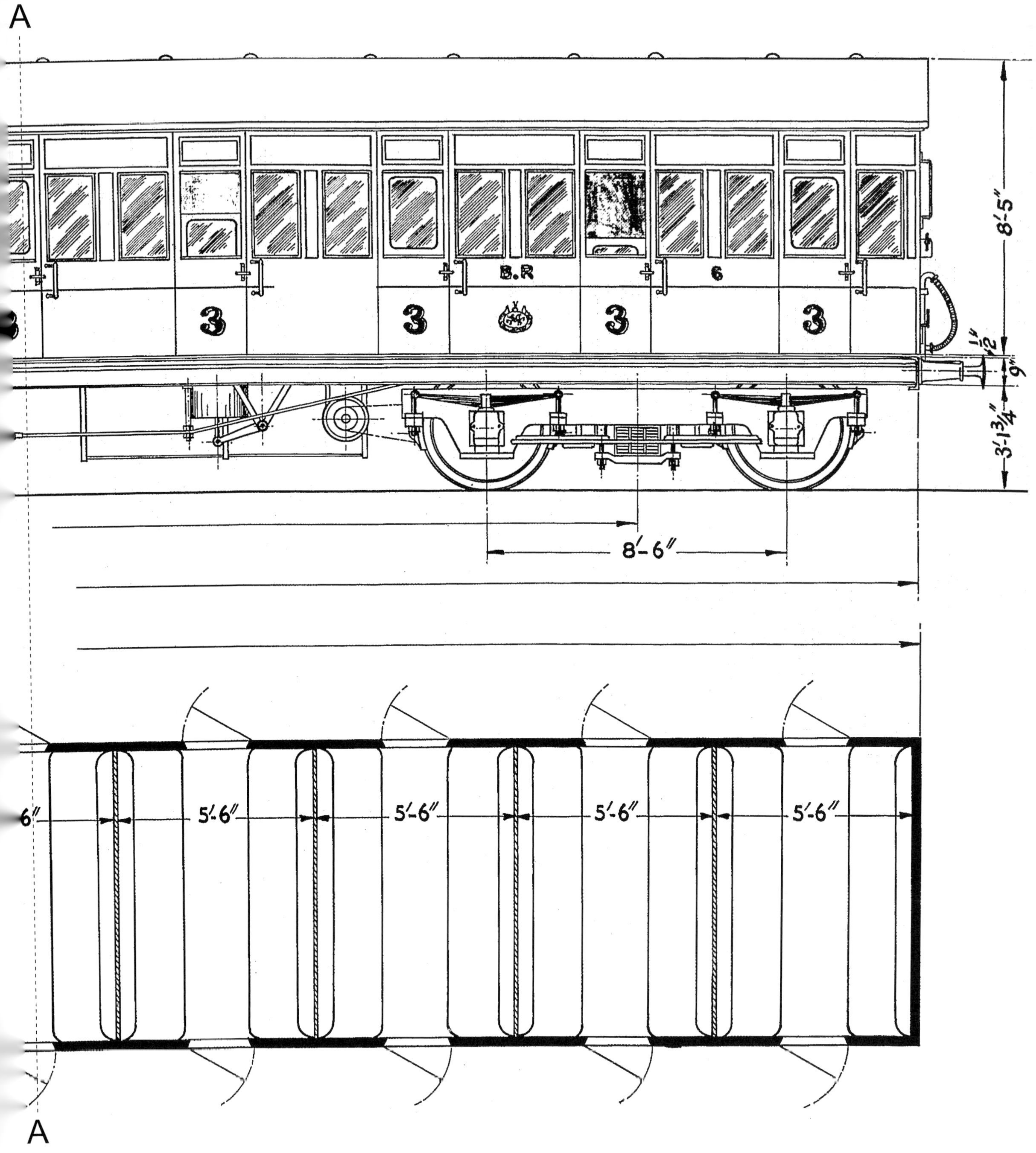

Six of these carriages were built by the Birmingham Railway Carriage & Wagon Company in 1920 as part of the seven-coach commuter set, the "Marble Arch" set. They had nine Third class compartments. The doors were fitted with Maxwell's Full Slam Locks, Meade & McClean Ventilators and the compartments fitted with Stone's electric lighting.

Drawing BRC6 Tricomposite Carriages Nos 150, 151, 152 and 156 (new series)

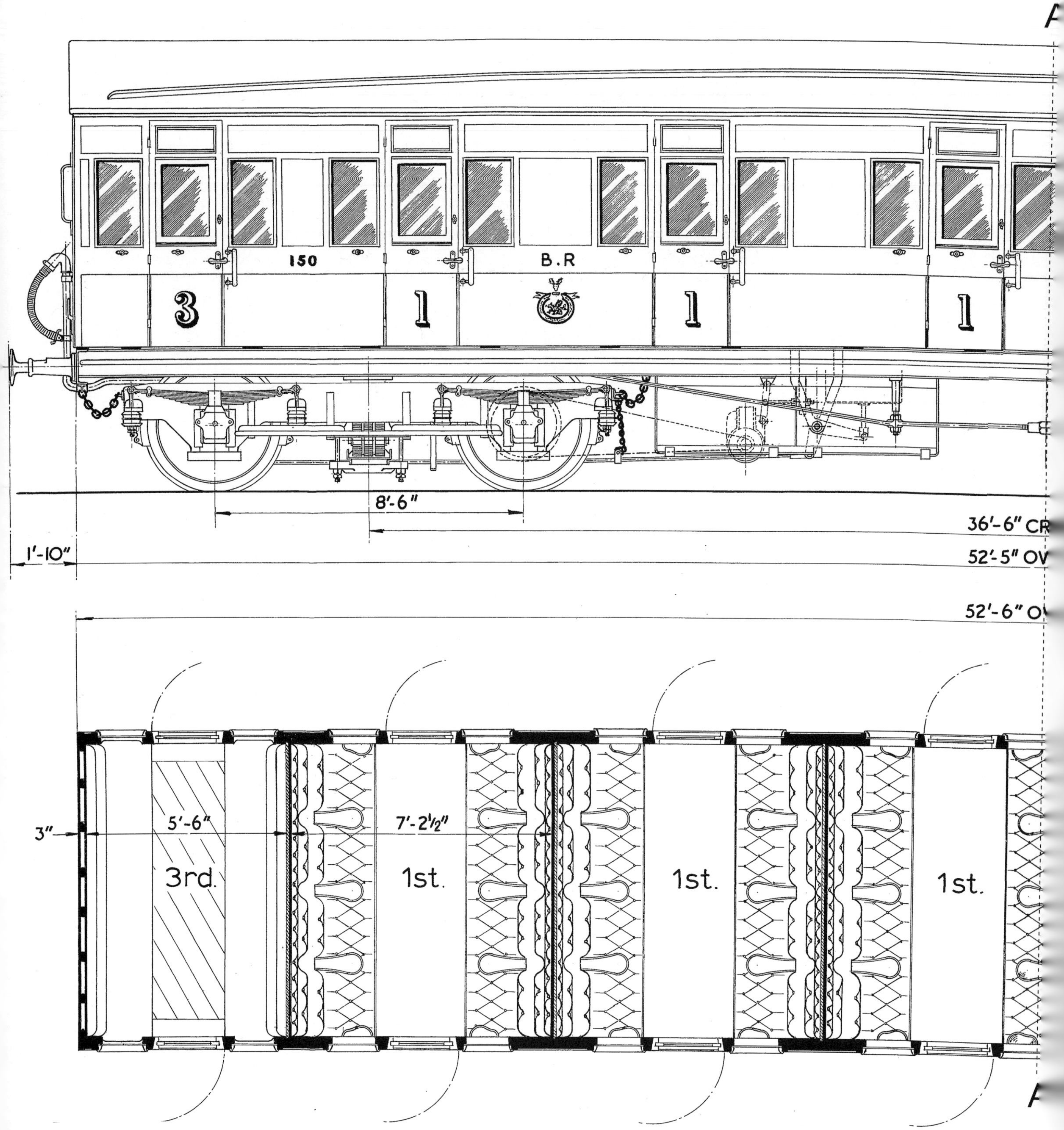

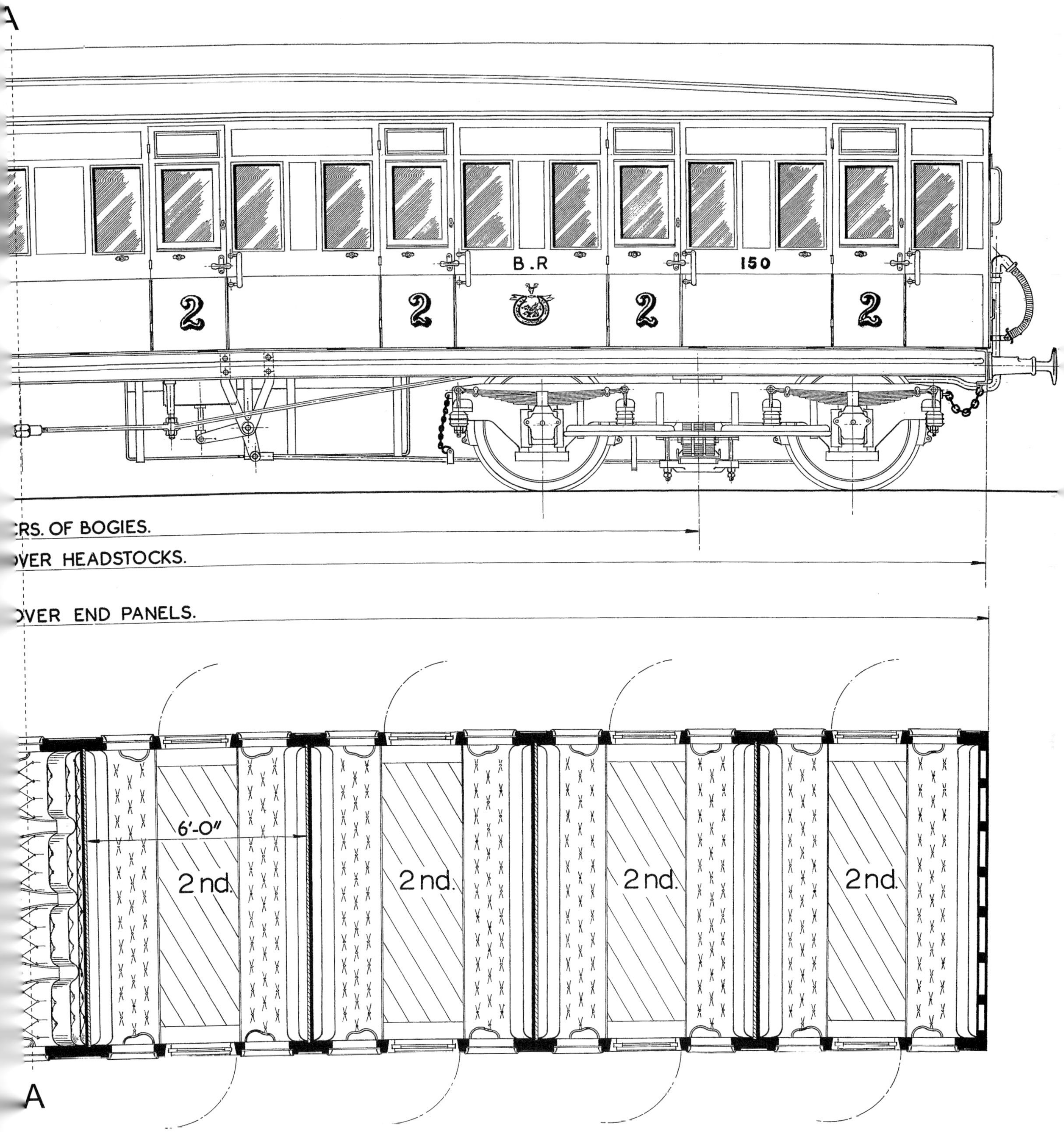
A
B.R
150
2
2
2
2
:RS. OF BOGIES.
OVER HEADSTOCKS.
OVER END PANELS.
6'-0"
2nd.
2nd.
2nd.
2nd.
A

Drawing BRC7 Brake Third Carriages Nos 197-200 (new series)

For end elevation, see page 100.

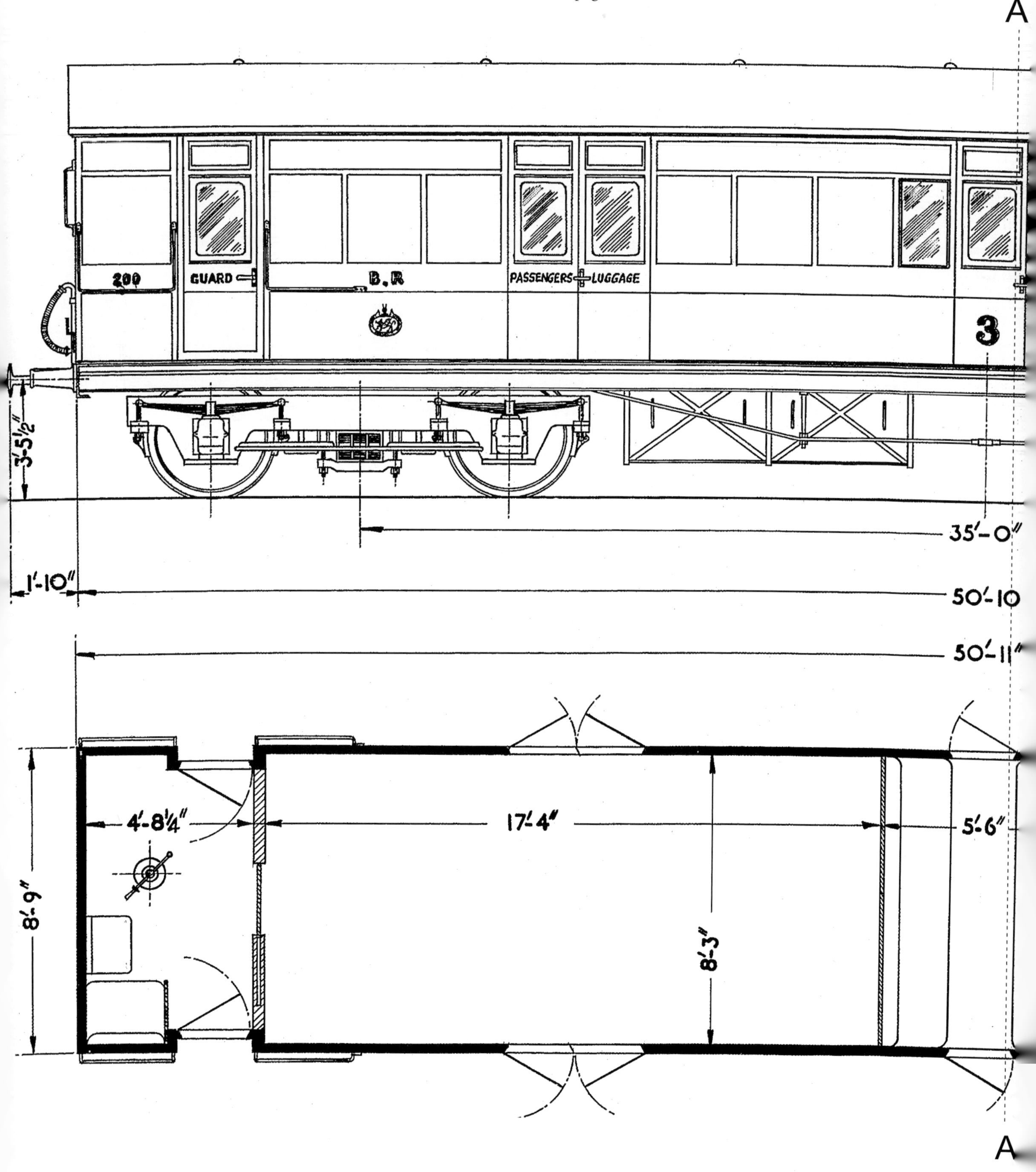

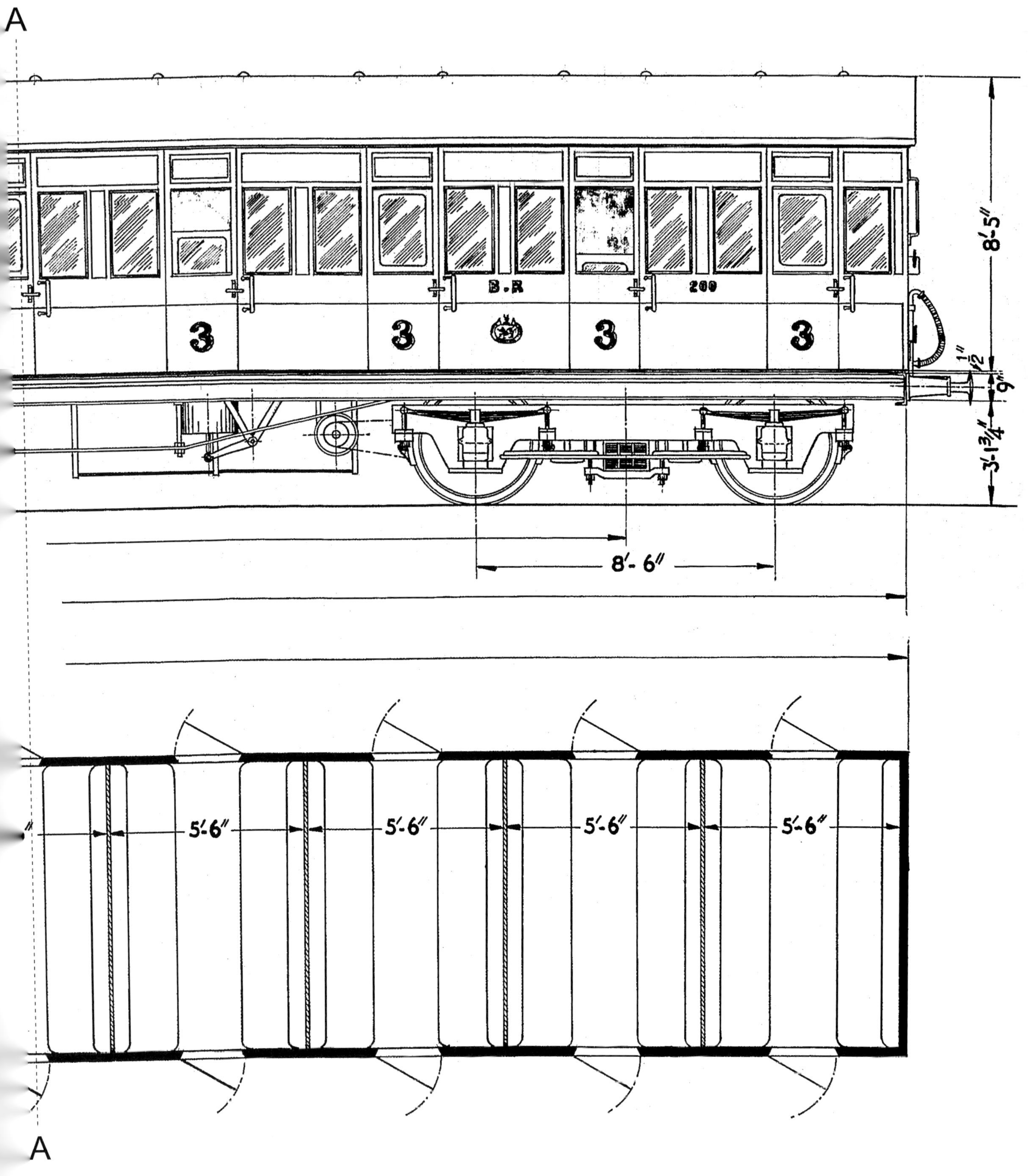
A
B.R
200
3
3
3
3
8'-5"
1"
½
9"
3'-1¾"
8'- 6"
5'-6"
5'-6"
5'-6"
5'-6"
A

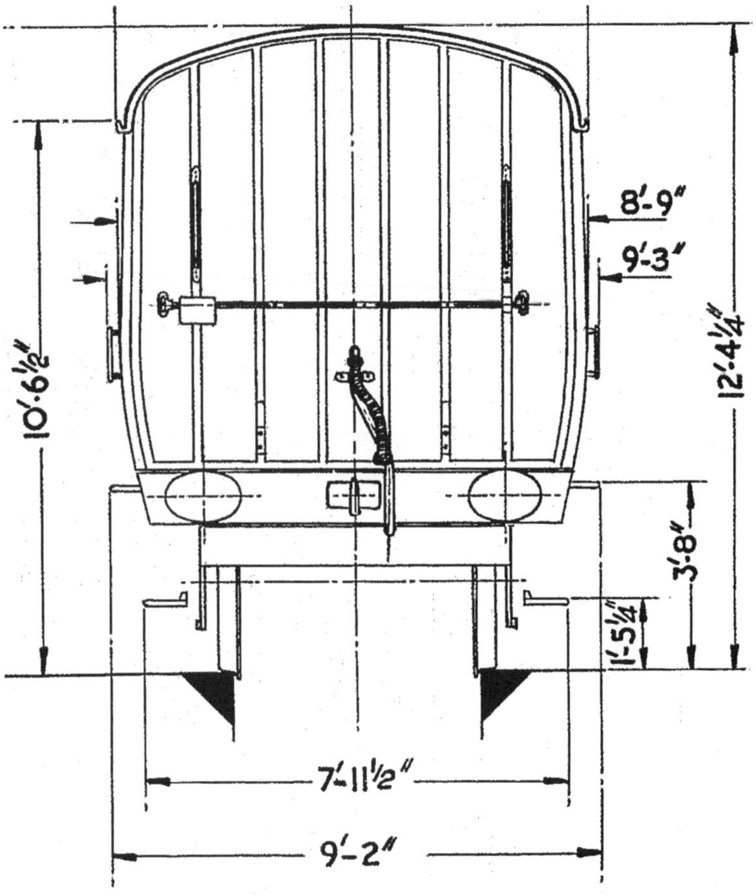

Looking somewhat the worse for wear, a "Marble Arch" Tricomposite is seen in Great Western livery at Wolverhampton. The four coaches of this type survived into nationalisation, the last of them being withdrawn on 25th April 1959. Note the transverse spring arrangement of the bogies. PHOTOGRAPHER UNKNOWN

Drawing BRC8: Bogie used on 1920 carriages

Scale: 14 mm/ft

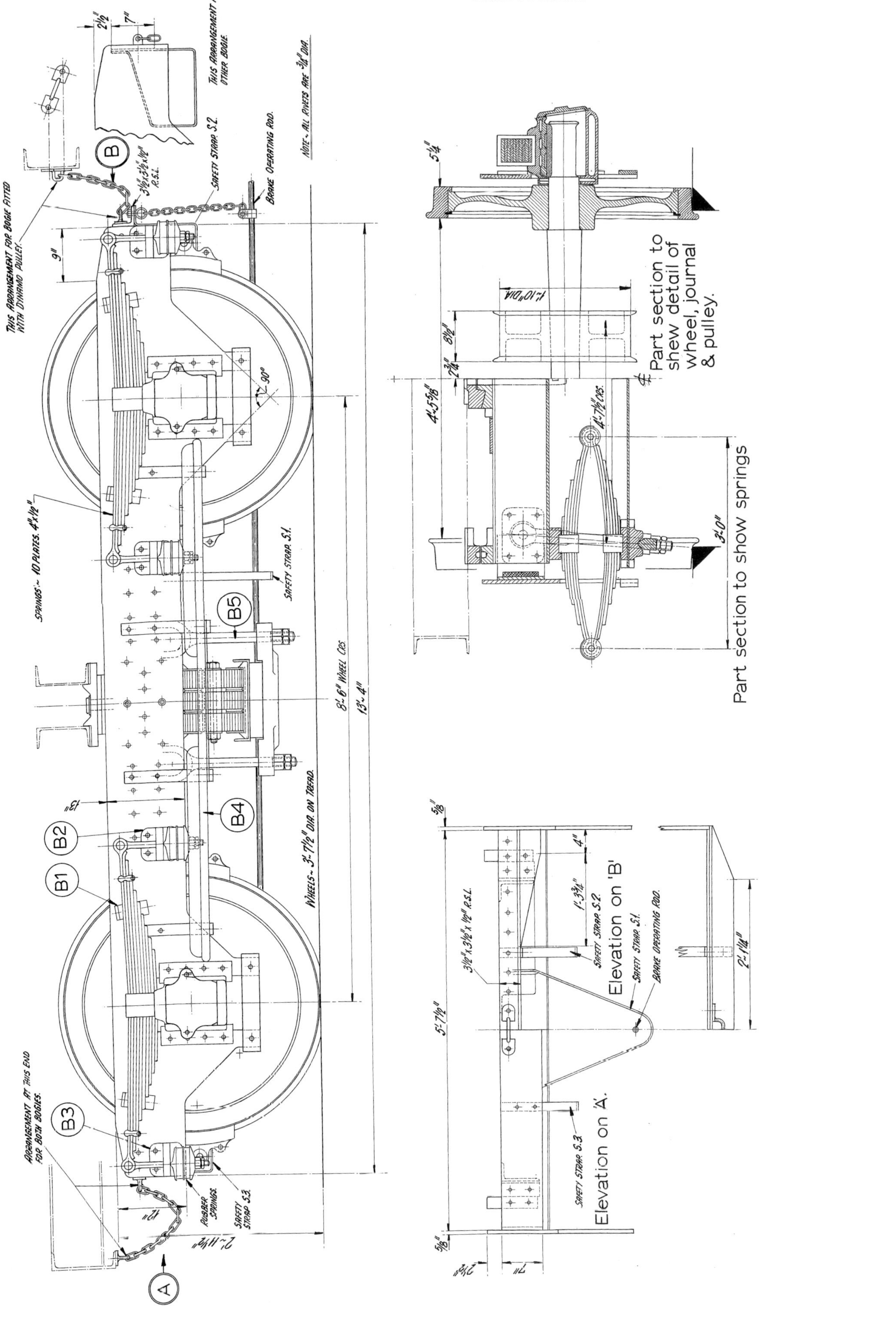

Drawing BRC9: Details for 8 ft 6 in. bogie used on 1920 stock

Scales:
Layout of brake linkage and detail of stepboards to bogie:
14 mm/ft
Details B1, B2, B3, B4 and B5:
28 mm/ft

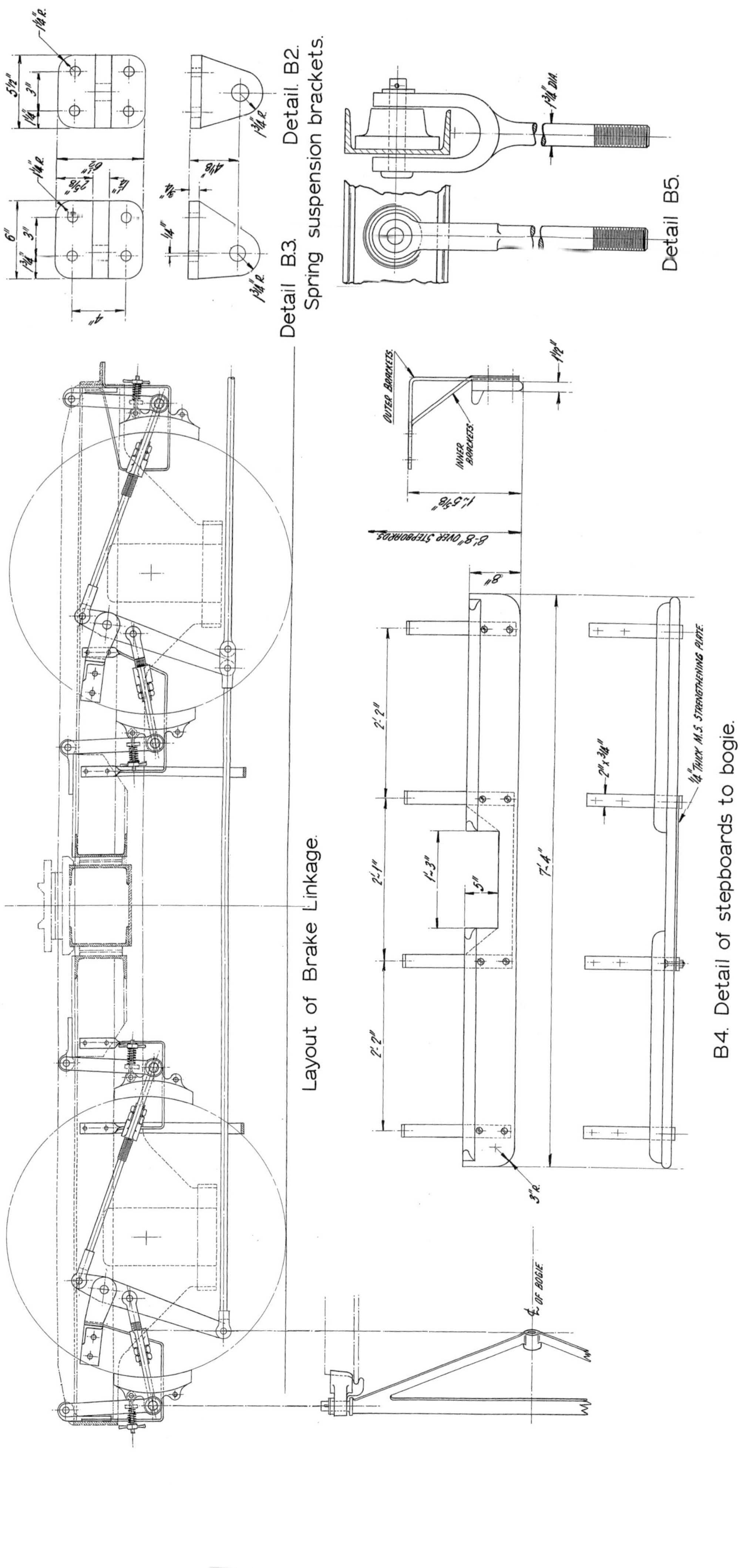

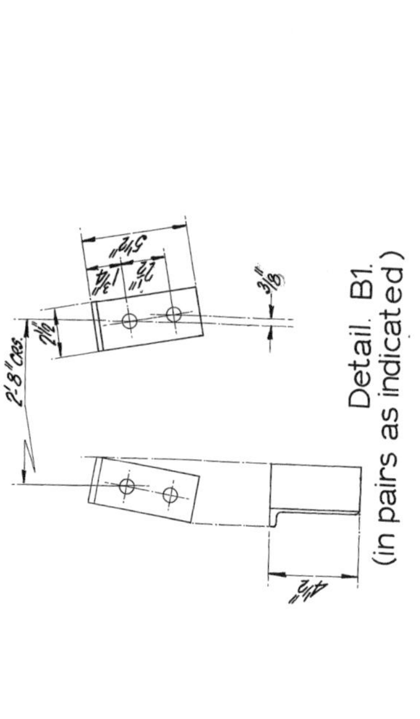

Drawing BRC10: Detail of Guard's Compartment, 1920 stock

Scale: 14 mm/ft

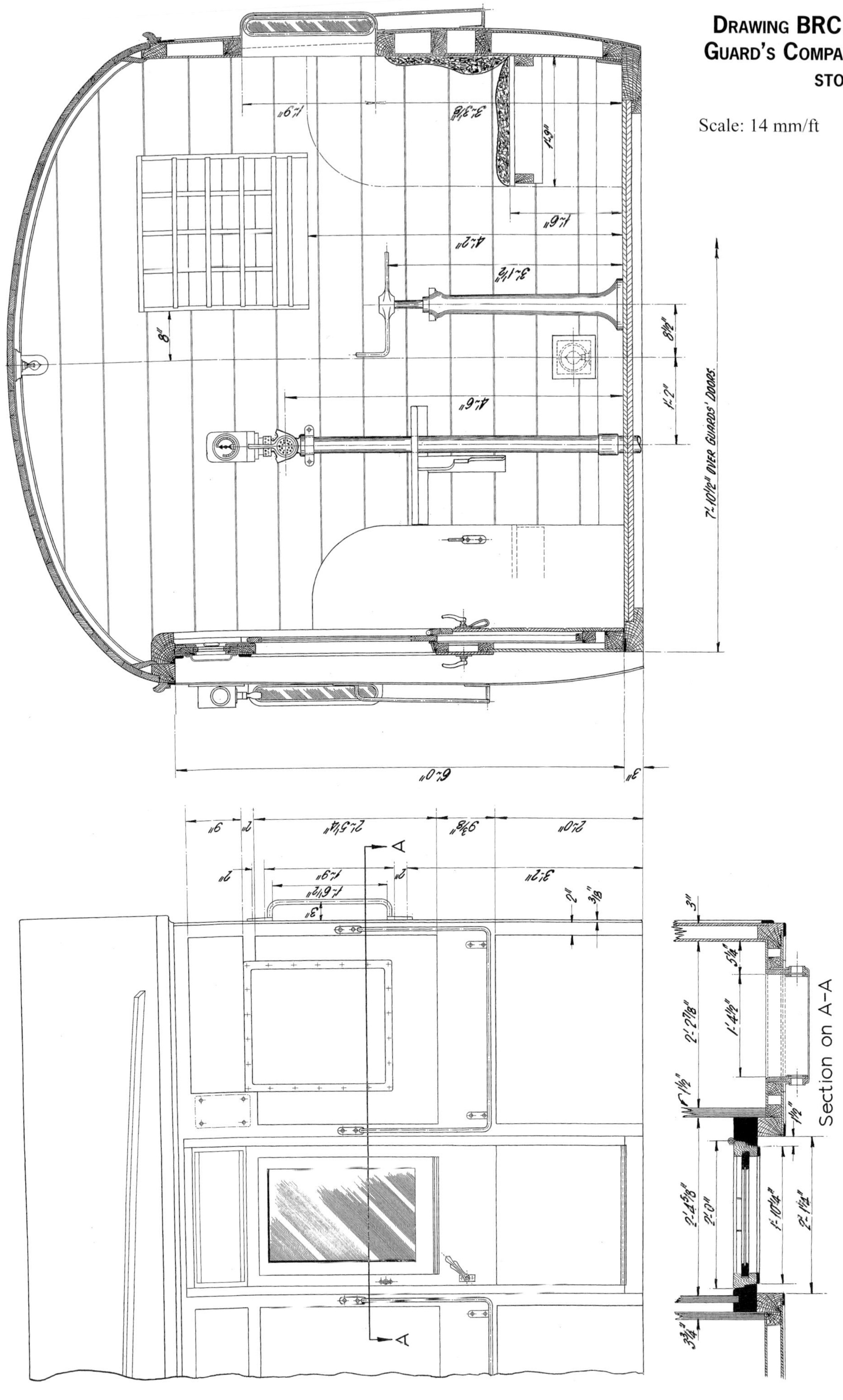

Drawing BRC11: End Elevation and Cross-Sections of First and Second Class Compartments for 1920 Carriage Stock built by the Birmingham Railway Carriage & Wagon Company

Scale: 14 mm/ft other than detail drawings.

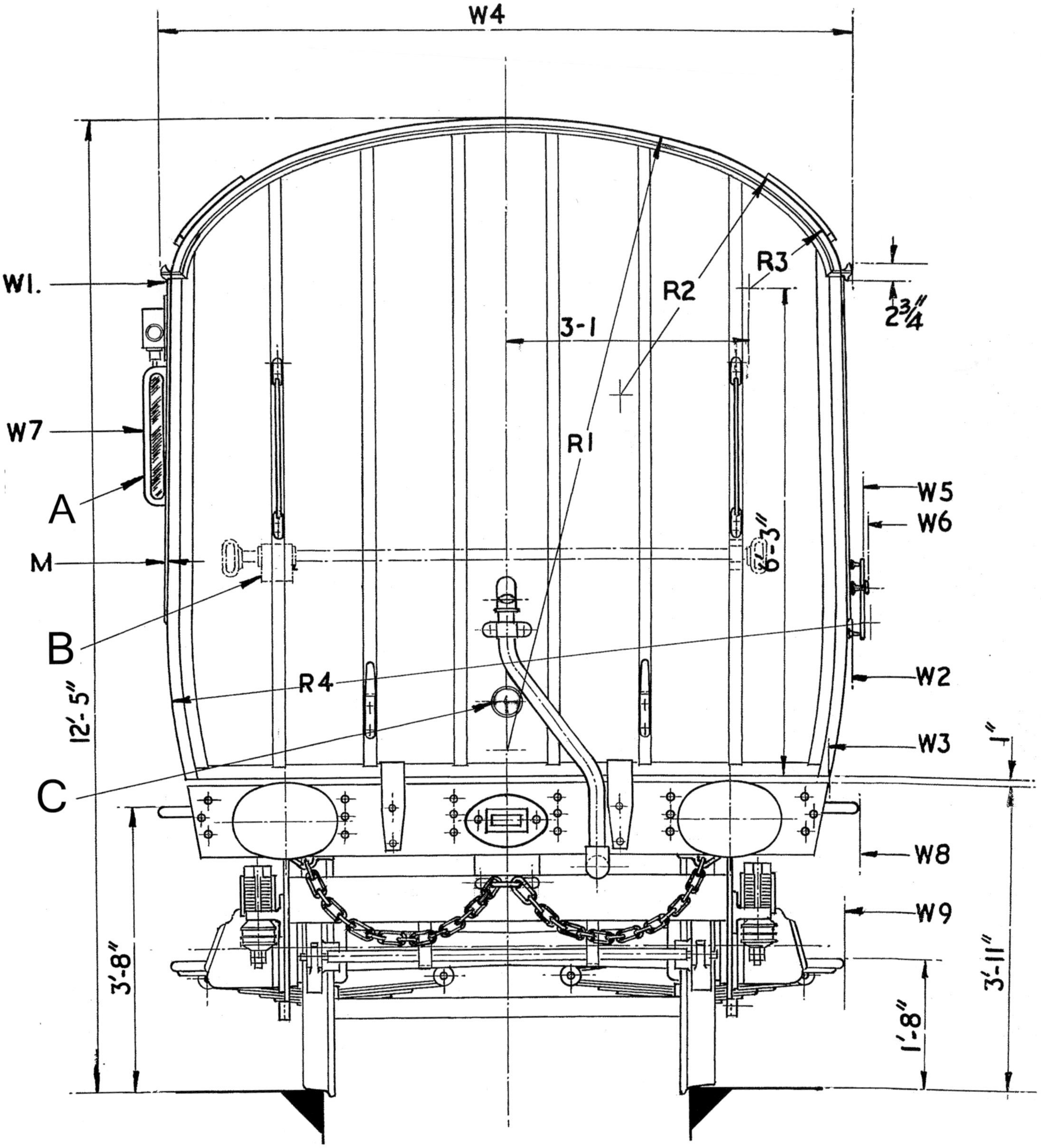

Notes and Key

A Both sides of Brake Third
B Location LS
C In Van end of Brake Third only
W1 8 ft 7 in. – width over mouldings at eaves
W2 8 ft 9 in. – width over mouldings at waist
W3 8 ft 3¼ in. – width over panels at bottom
W4 8 ft 9¾ in. – width over gutter mouldings
W5 9 ft 1$^{9}/_{16}$ in. – width over commode handle
W6 9 ft 2⅜ in. – width over door handles
W7 9 ft 3 in. – width over Guard's "lookouts"
W8 9 ft 0 in. – width over top step boards
W9 8 ft 8 in. – width over lower step boards
M ⅜ in. – thickness of mouldings
R1 8 ft 1 in. – radius of major arc
R2 3 ft 4 in. – radius of blending arc
R3 1 ft 2½ in. – radius of minor arc
R4 9 ft 0 in. – radius of bodyside "tumblehome"

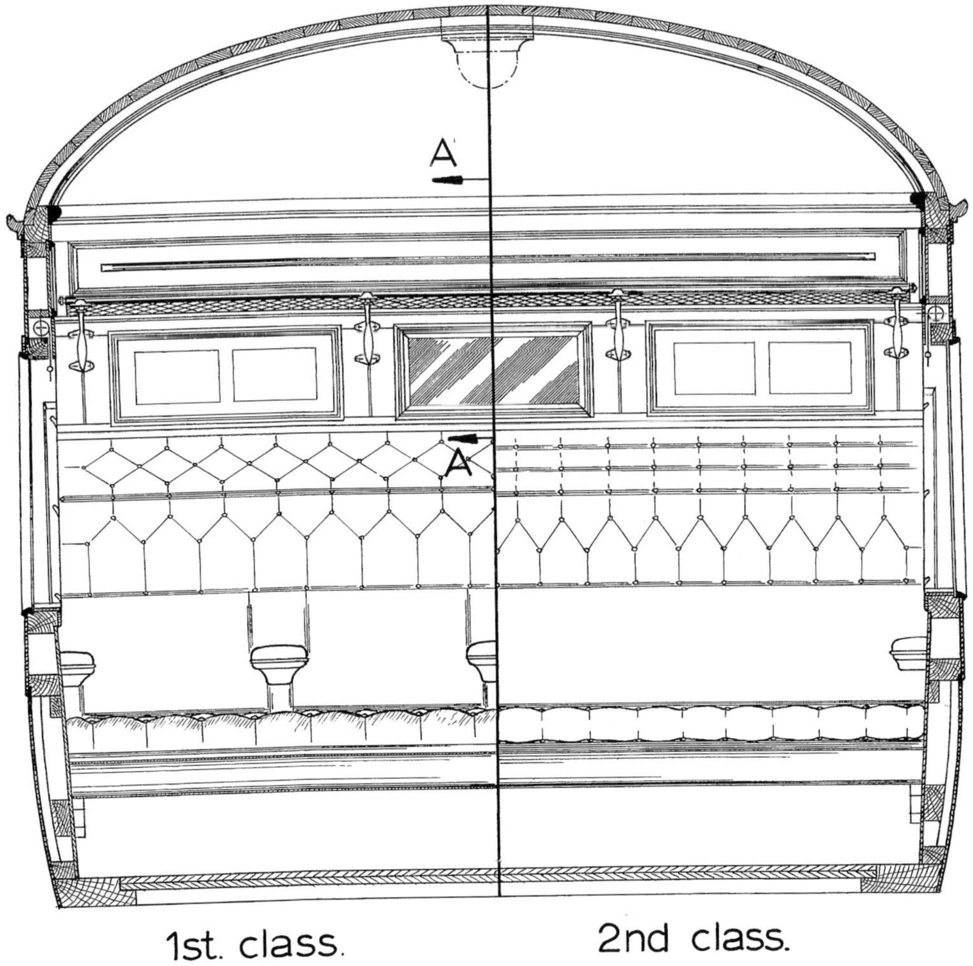

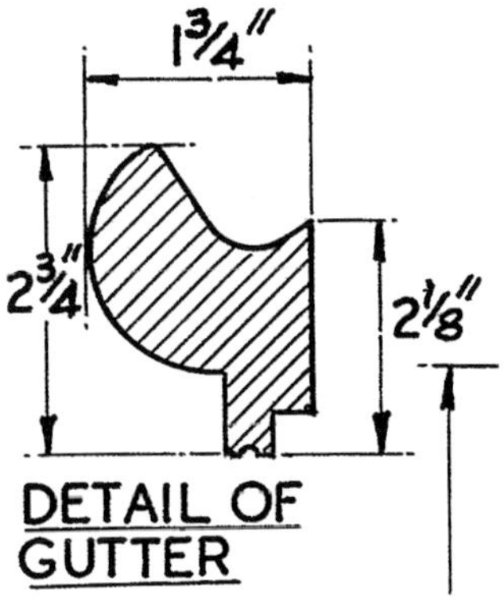

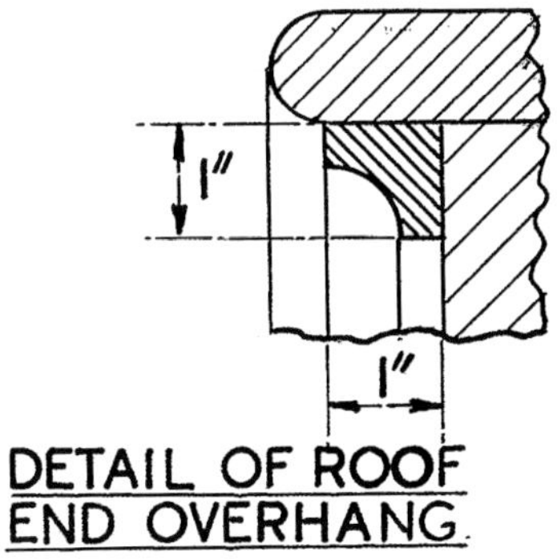

Note:
M1-M6 are not defined here or in drawing BRC4/1.

Section on A-A 20% of full size

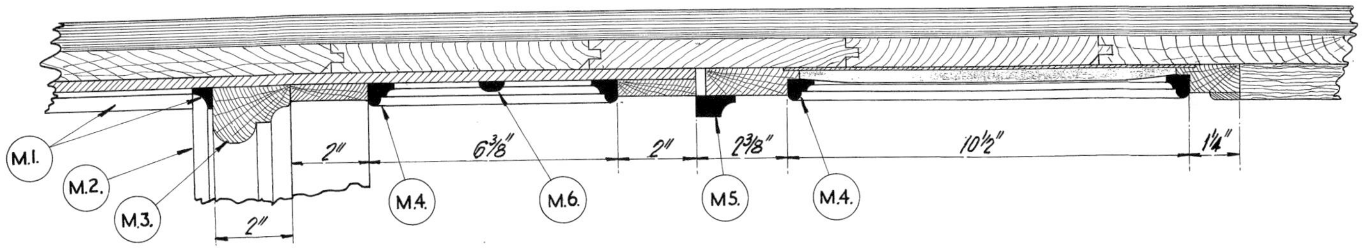

In 1920, in a statement of post-war optimism, the Barry Railway ordered new bogie coaches for the first time in its history. Six Thirds, four Tri-Composites and four Brake Thirds were supplied by the Birmingham Railway Carriage & Wagon Co. This provided the Barry Company with two sets of seven coaches. They were known as the "Marble Arch" sets owing to the high arch of the coach roofs, as can be seen from the photographs and drawings. Their primary use seems to have been on rush hour commuter traffic between Llantwit Major and Cardiff. PHOTOGRAPHER UNKNOWN

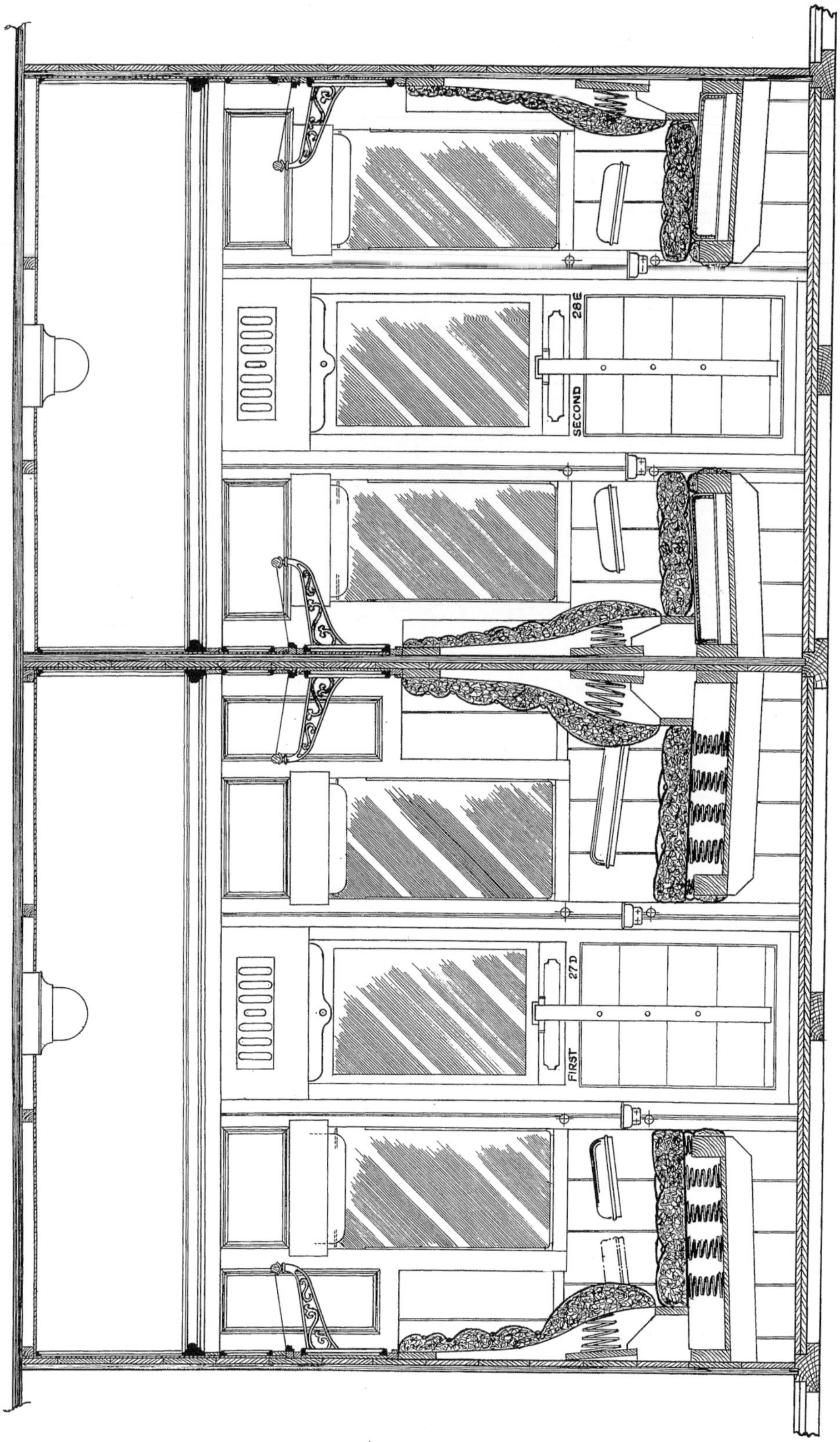

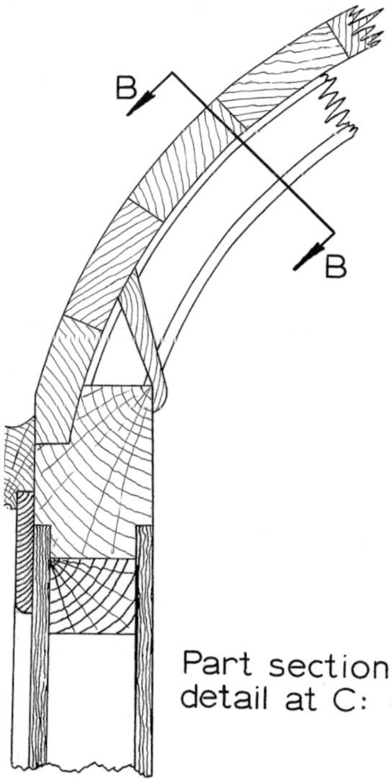

Part section detail at C:

Left: *First Class (upper) and Second Class (lower)*

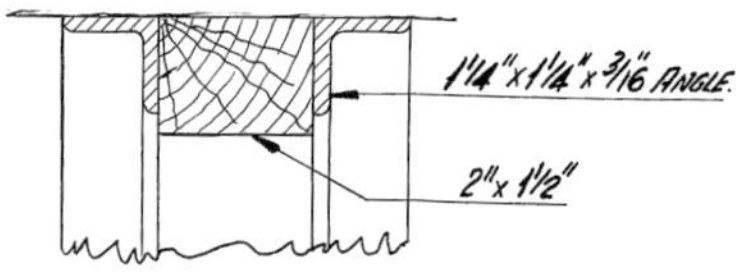

Section on B-B shewing detail of roof rib at R1.

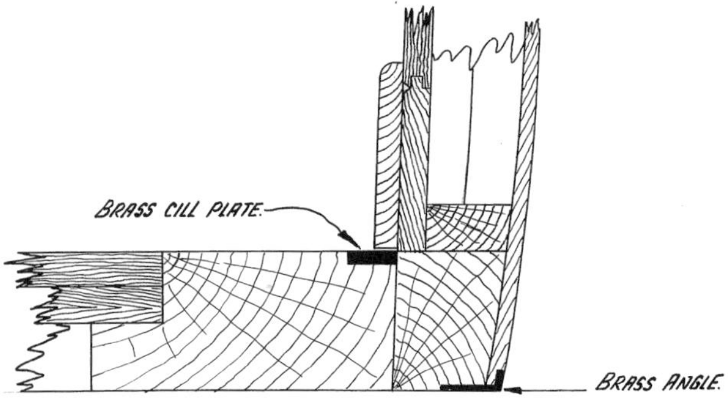

Part section shewing detail at D: half size.

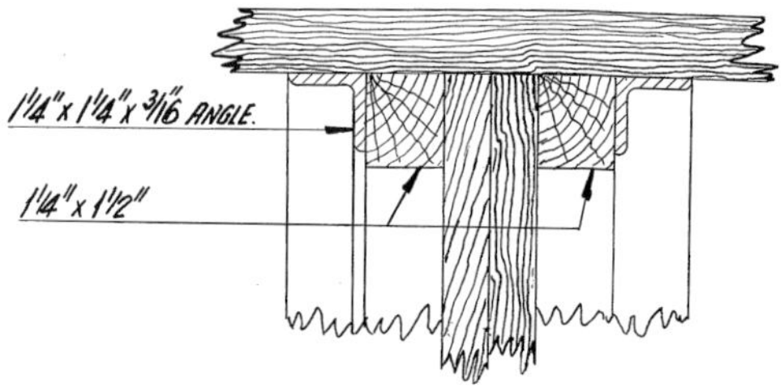

Section on B-B shewing detail of roof rib at R2.

Drawing BRC12: Details of Interiors of Compartments for 1920 Carriage Stock built by the Birmingham Railway Carriage & Wagon Company

Top: *Third Class compartment*

Bottom: *Workmen's compartment, section on longitudinal centreline and cross-section*

Scale: 14 mm/ft for cross-section; 20% of full size for part sections; detail drawings not scaled

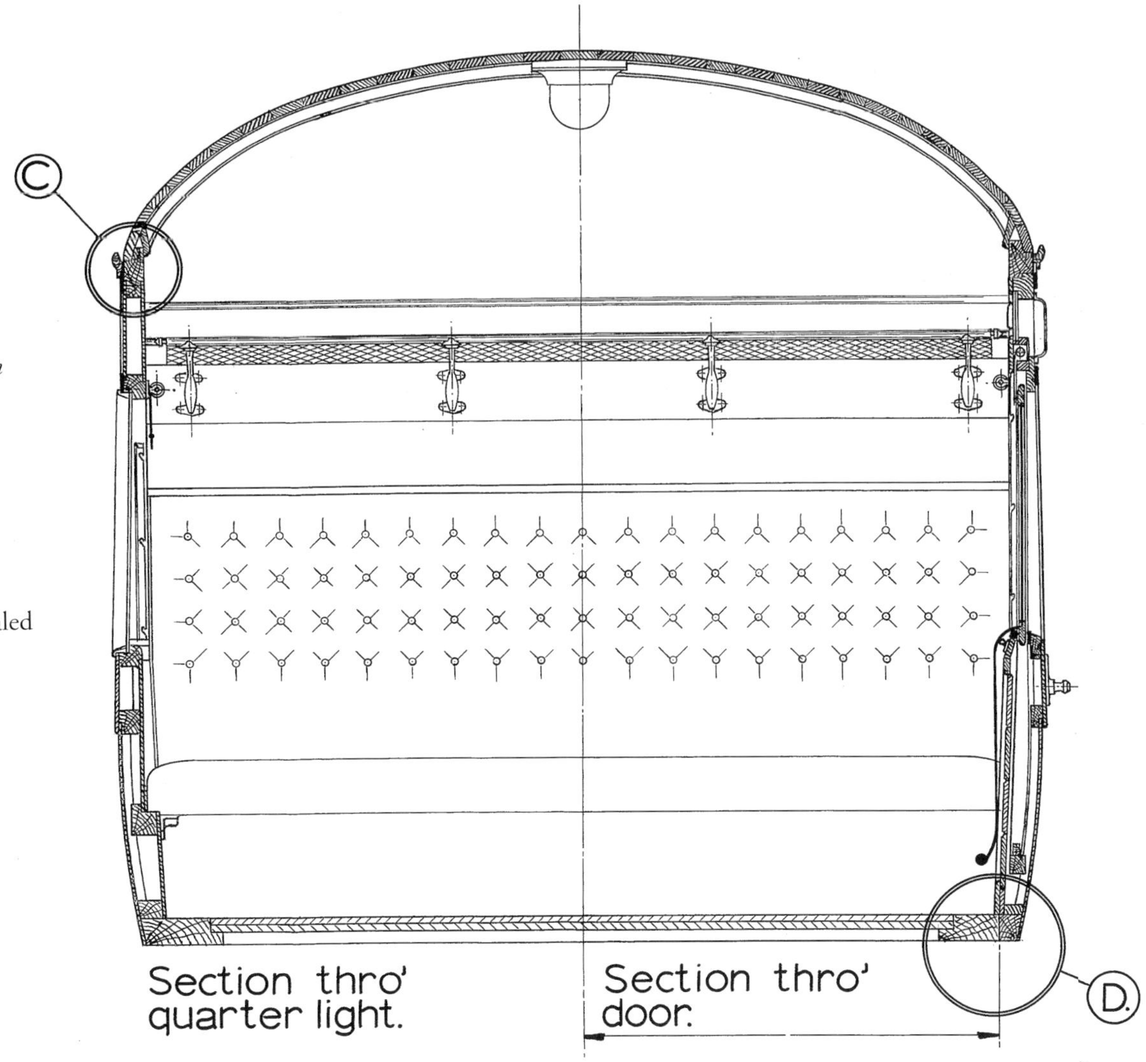

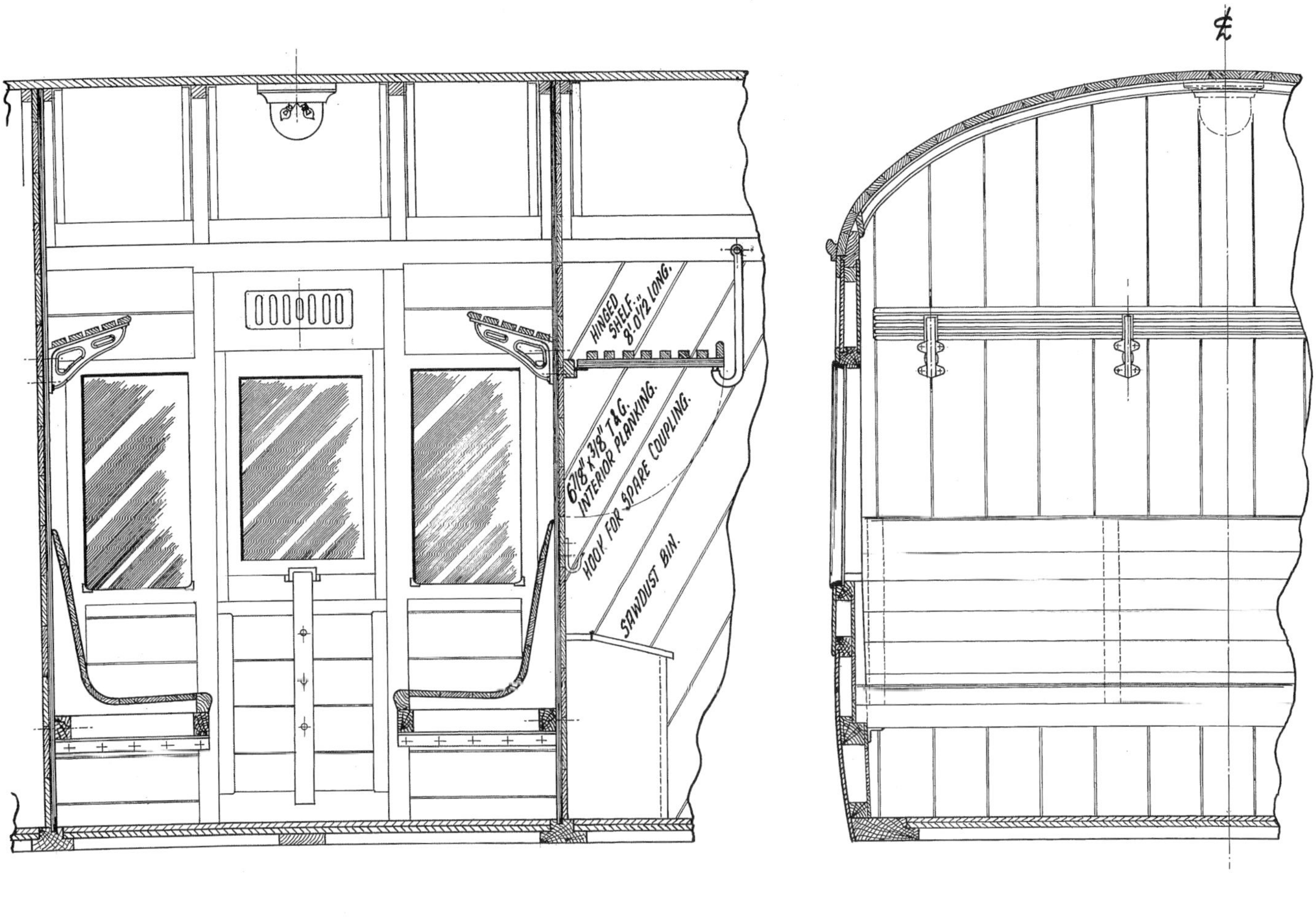

Drawing BRC13: Detail of carriage side, Tricomposite, 1920 stock

Scale: 14 mm/ft

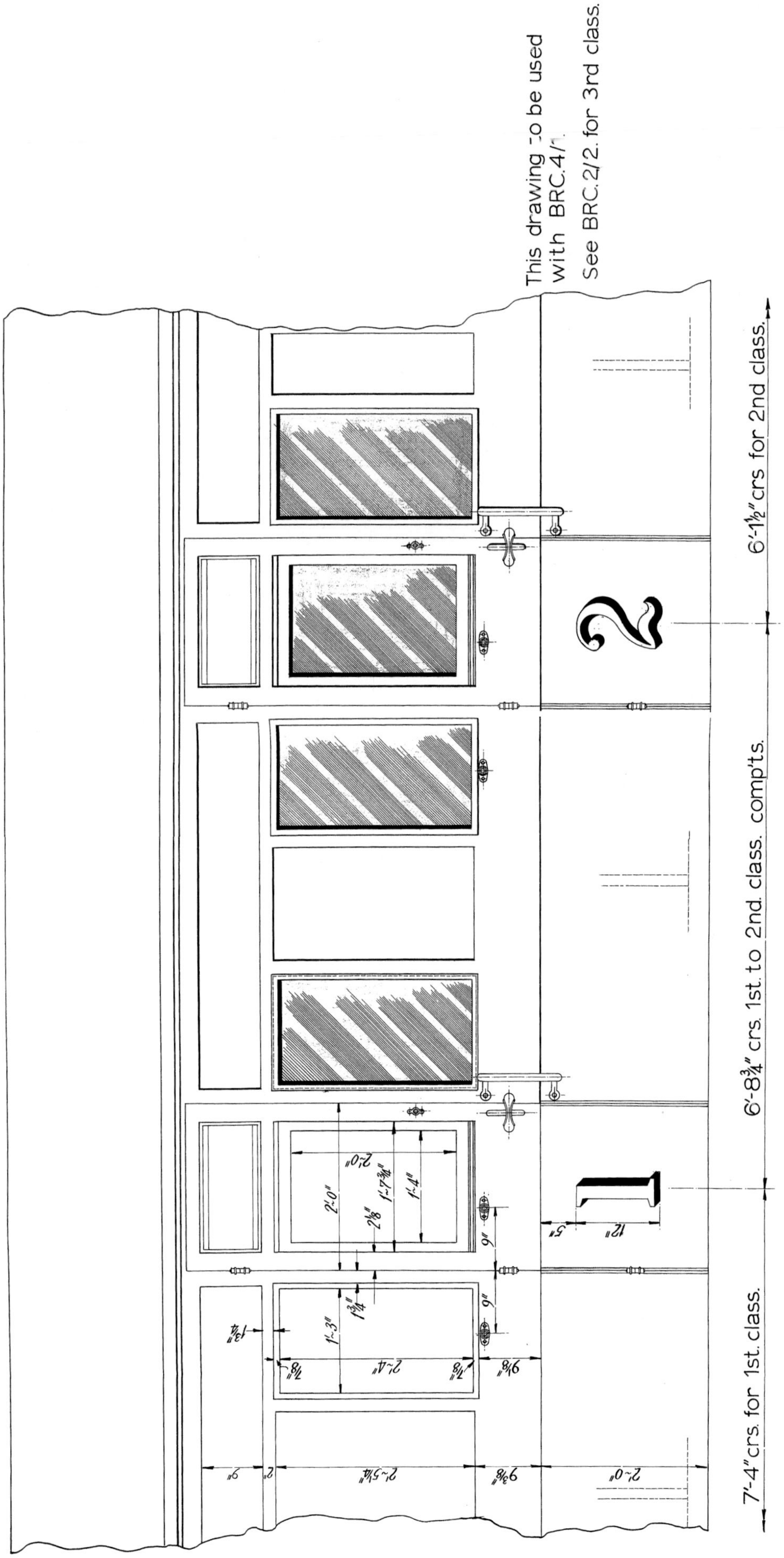

Drawing BRC14: Detail of carriage side and Longitudinal Section through Third Class compartment, 1920 stock

Scale: 14 mm/ft

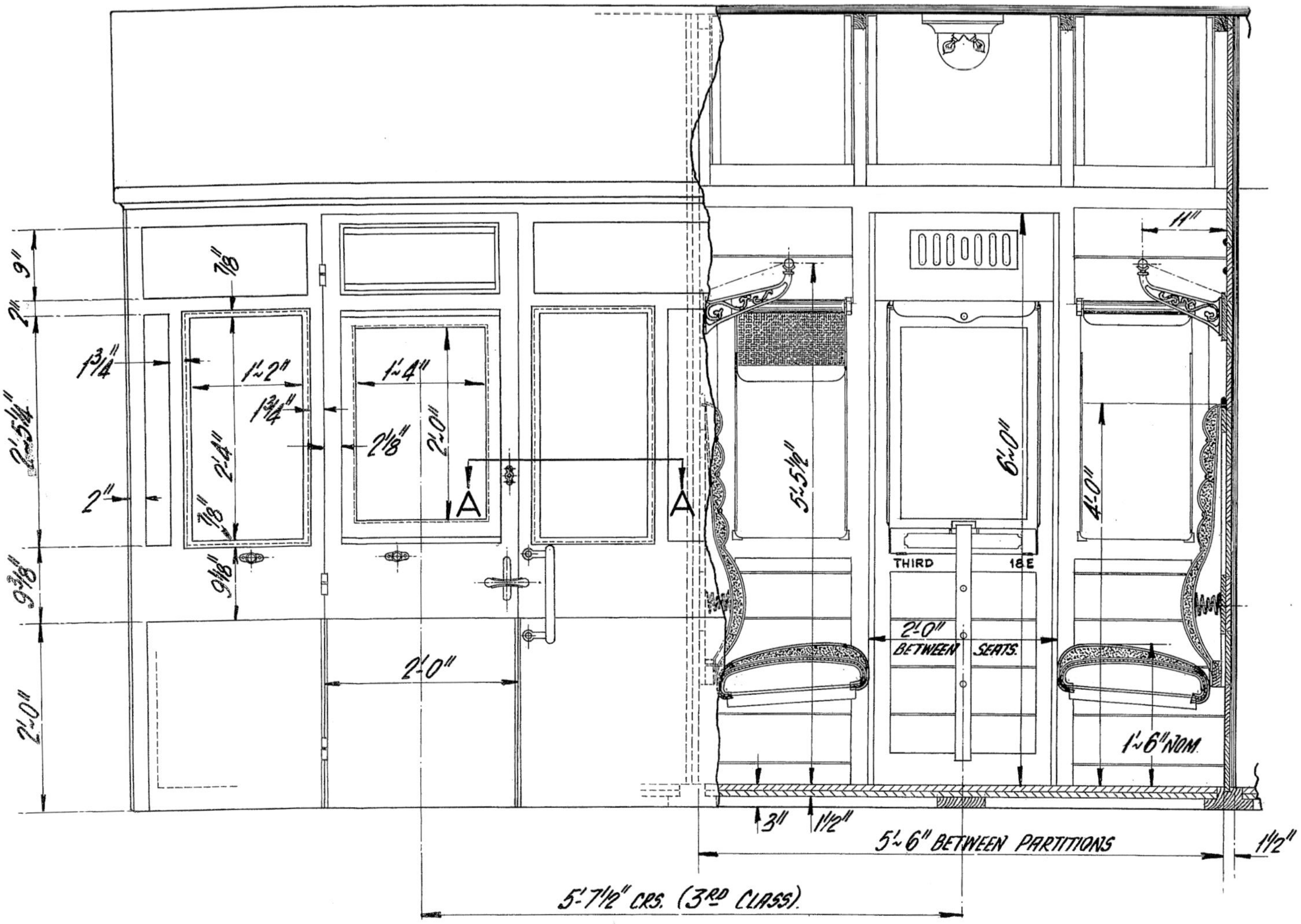

Section on A-A, scale 20% of full size

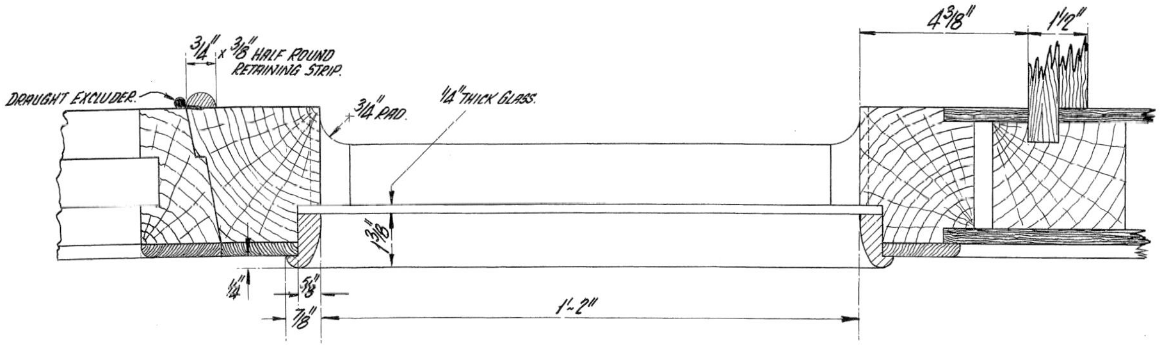

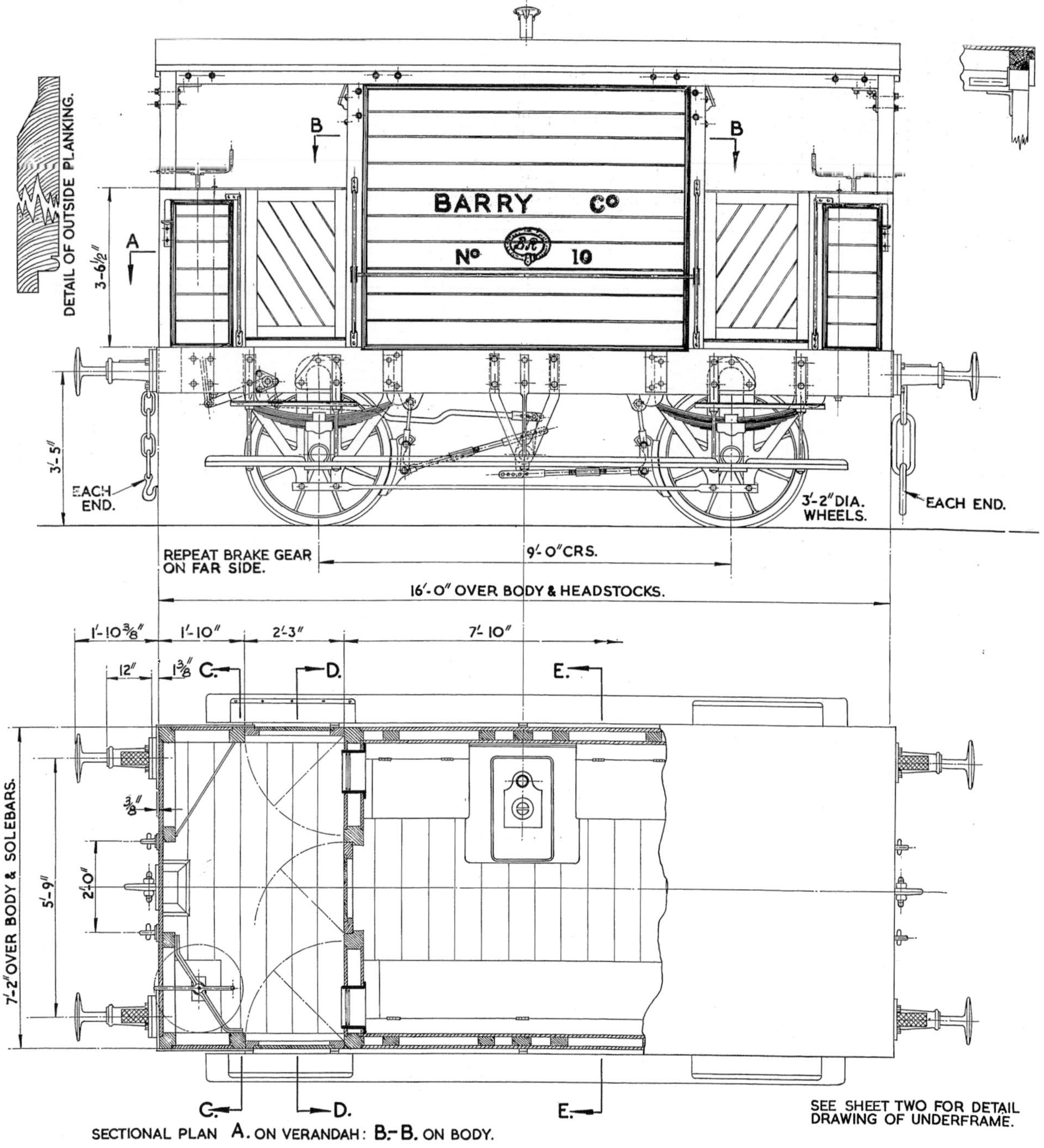

Drawing BRB1 Four-wheeled Brake Vans Nos 1-61

Although ordered over a period of 12 years, all 61 four-wheeled mineral brake vans obtained by the company were essentially similar. The main difference was that 21 were of only 7 tons tare weight, the rest being 11 tons. Each batch ordered contained both variants. The first 15 were obtained in 1889 from Metropolitan but the next batch of ten came from the Swansea Wagon Co in 1890. The same year the Metropolitan supplied a further five, while in 1896 15 were supplied by the Midland Wagon Co. Finally, in 1900 15 more were obtained from Ashbury. One difference between batches was that the first three had laminated buffers, the latter two "rubber" buffers. All were vacuum piped according to Mountford, though neither drawings nor photographs show this so there must be considerable doubt. Both No. 10 and No. 47 were withdrawn in January 1926.

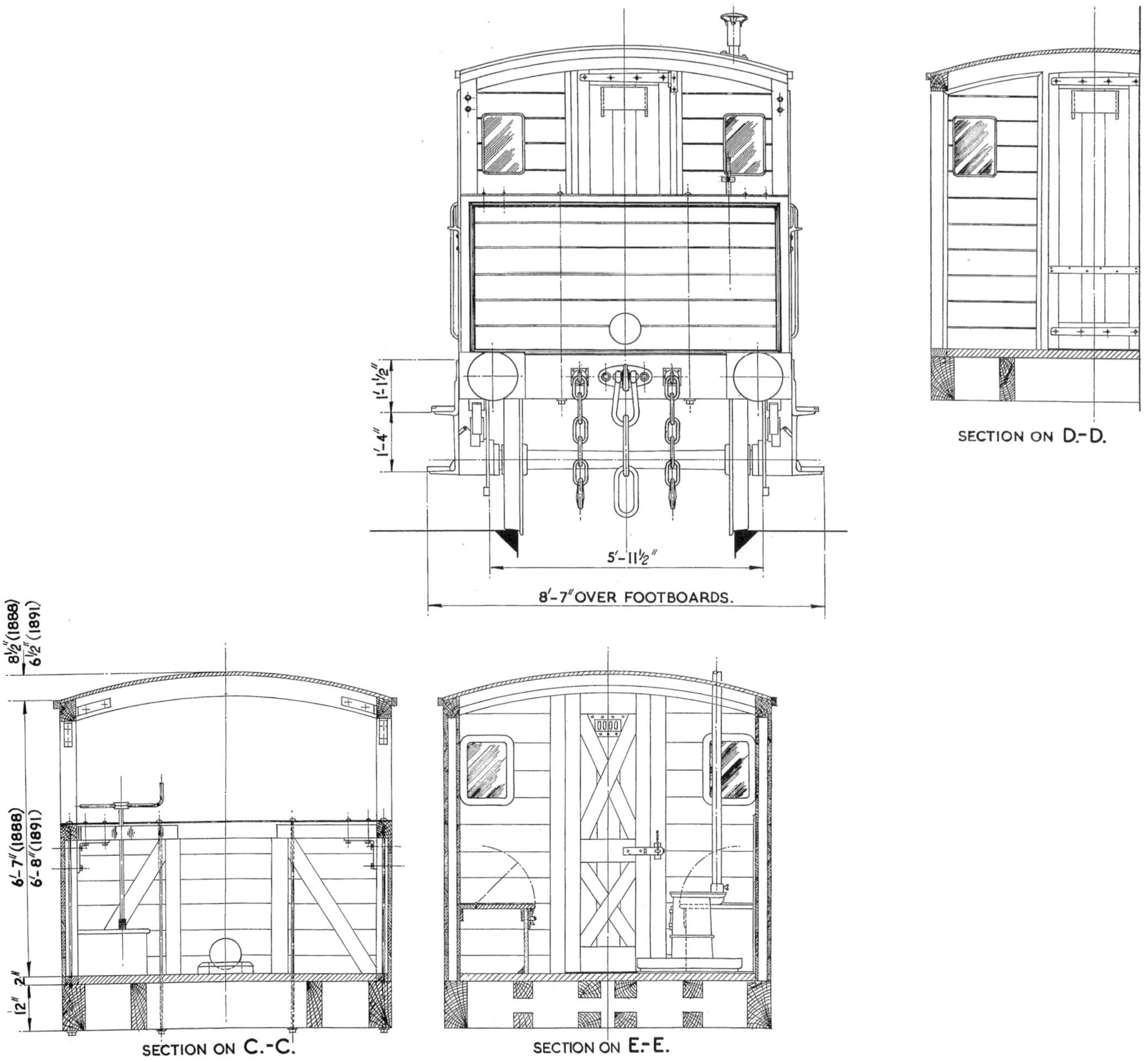

Ashbury Wagon Co., Gregory collection

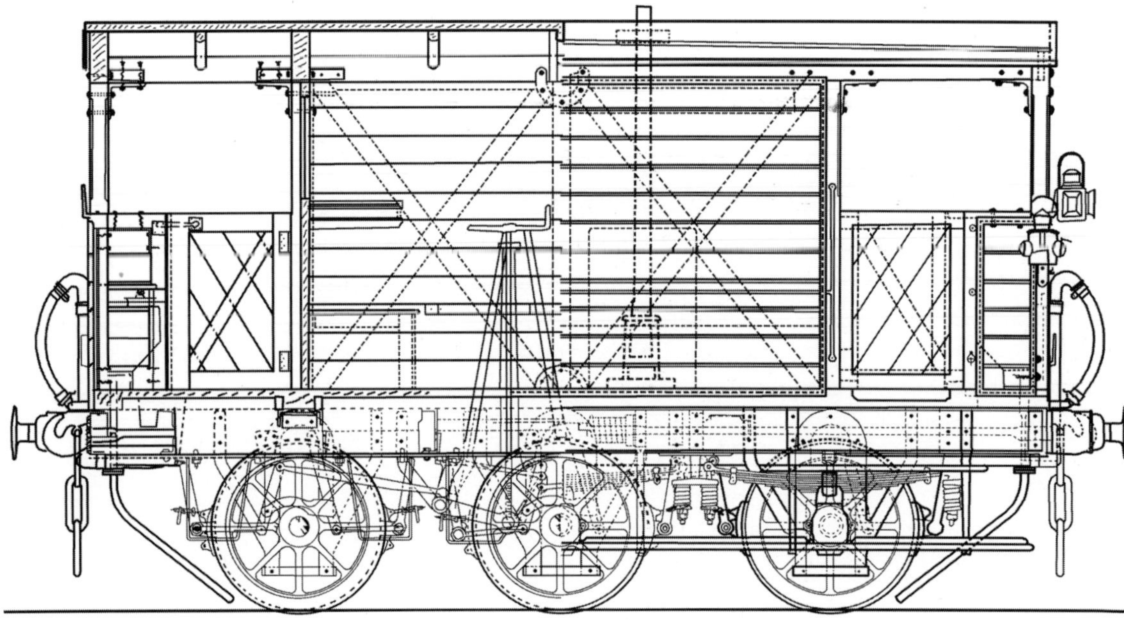

Redrawn by Jonathan David

Drawing BRB2
Six-wheeled Brake Vans Nos 62–73

Until 1915 all Barry Railway mineral brake vans were of essentially the same four-wheeled design. In that year 12 heavy six-wheeled vans were obtained from the Midland Carriage & Wagon Co. These turned out to be the last brake vans obtained by the company. They were definitely vacuum piped. At 20 tons tare rather than 7 or 10 tons they were a significant advance on all the previous Barry Railway brake vans. All 12 lasted until after nationalisation.

These drawings have been prepared for this book from official General Arrangement drawing 55-31 (though on the original drawing this has been crossed out with 34-270 written underneath).

Overall length: 21 ft 0 in.
Length over headstocks: 18 ft 0 in.
Wheelbase: 5 ft 0 in. + 5 ft 0 in.
Length of lower footboards: 17 ft 7 in.
Overall height: 11 ft 0½ in.
Solebars and buffer beams: 10 in. deep
Width over handrails: 8 ft 1½ in.
Width over lower steps: 8 ft 8 in.
Roof radius: 11 ft 0 in.
Bearing springs: eight plates 4 in. × ½ in.
Upper stepboards: 4 ft 6 in.
Platforms: 3 ft 6 in. by 6 ft 10 in. clear
Floor boards 1½ in. thick, tongue and groove

Upper lockers on the plan view are for coal; the lower left hand locker is for lamps, the others for general storage.
Lamps to match samples supplied by the Barry Railway Co.
Position of Barry Railway Co name and number plate to be on left hand side when facing view (so on the far side on this drawing).
Fold-up seats over all four sandboxes.
Lead sheet in entrances fixed with copper nails.
Note that the through vacuum pipe is offset and does not change sides under the vehicle.

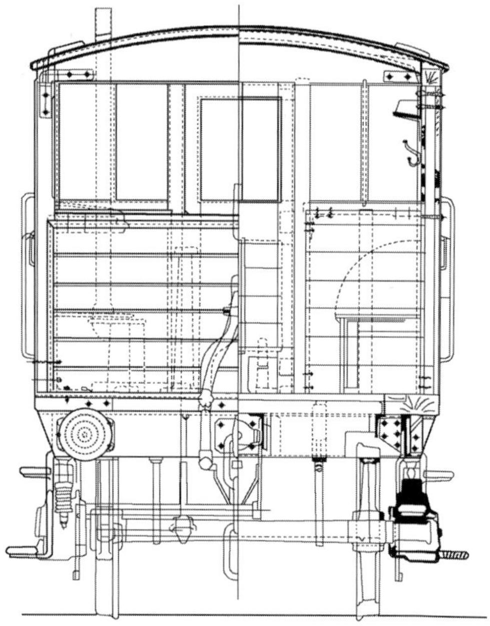

Locomotive and Mechanical Committee
2 May 1919
Breakdown crane. Referring to Minutes of the last Traffic Committee, the General Manager submitted an offer from Messrs Cohen & Sheldon of Carlisle* to supply a crane to the Company's specification for £5,600, including truck for carrying the jib. Messrs Stothert & Pitt of Bath had also been asked to quote, but their offer had not been received.

The General Manager was authorised to accept the most satisfactory of the two offers up to a sum of £6,000.

* Almost certainly an error for Cowans & Sheldon

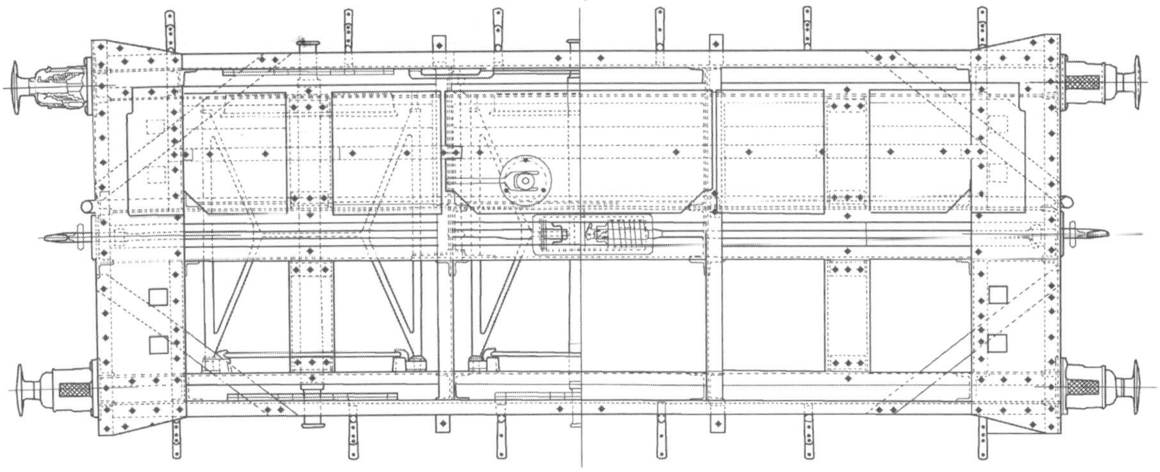

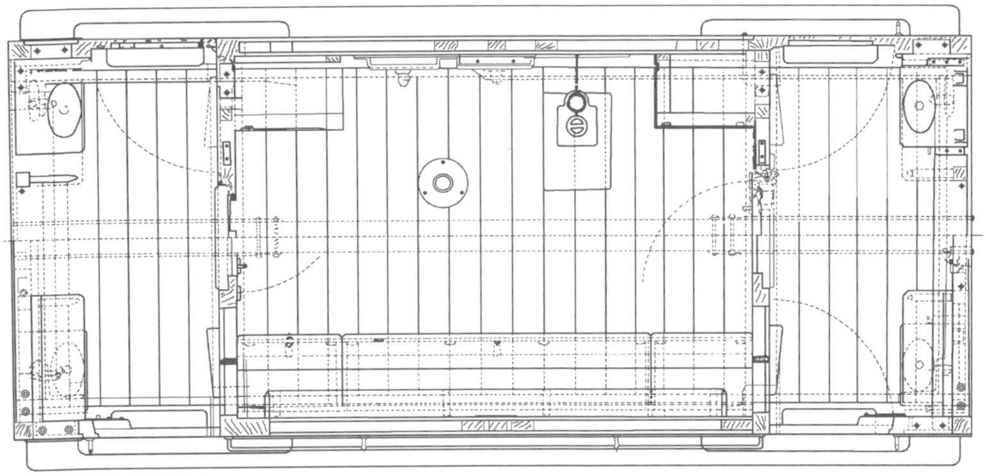

MIDLAND CARRIAGE & WAGON CO.

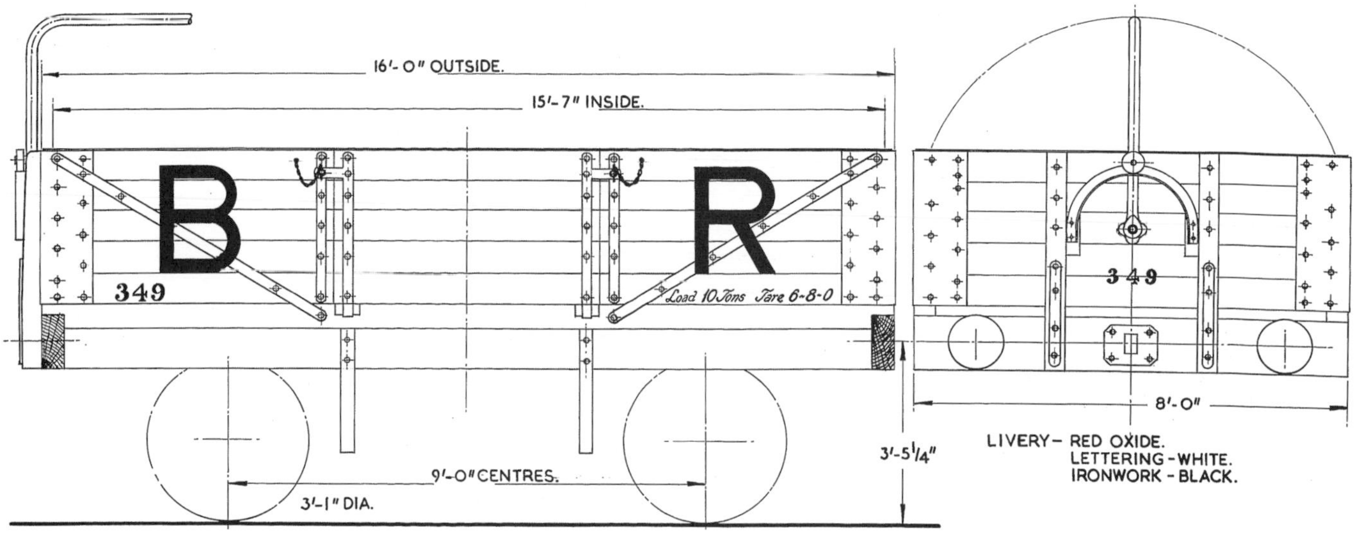

Drawing BRW1 Five Plank 10 Ton Open Goods Wagon with Tarpaulin Rail

One of a large batch of wagons, probably 150, built in 1920 by S J Claye but essentially the same as a batch built by the same company in 1912. All replaced withdrawn wagons and took their numbers. By this time the Board of Directors did not usually deal with rolling stock orders so we do not have details such as the price paid, though the General Manager had reported on the shortage of wagon stock and recommended that 200 additional open wagons be obtained. He was authorised to prepare drawings and obtain tenders. He submitted tenders later the same month and four months later was authorised to purchase them. No. 349 was withdrawn in December 1930.

Five plank open with tarpaulin rail No. 13, built by the Barry company in 1910. The 300 wagons built by Harrison & Camm between 1912 and 1920 were very similar; see drawing above. HMRS Jim Richards collection AAJ920

Drawing BRW2 Four Plank 10 Ton Open Goods Wagon

The Barry company built many wagons over the years to replace withdrawn vehicles. This example was built by the Barry Railway in 1903 as one of a batch of ten wagons. Essentially similar wagons were built in 1898, 1899, 1901, 1902, 1904, 1905 and 1907. See the Revenue Vehicles section of the Goods Rolling Stock List for details of differences between batches. Again these orders are not recorded in the minutes of Directors' meetings. No. 11 was withdrawn in May 1931.

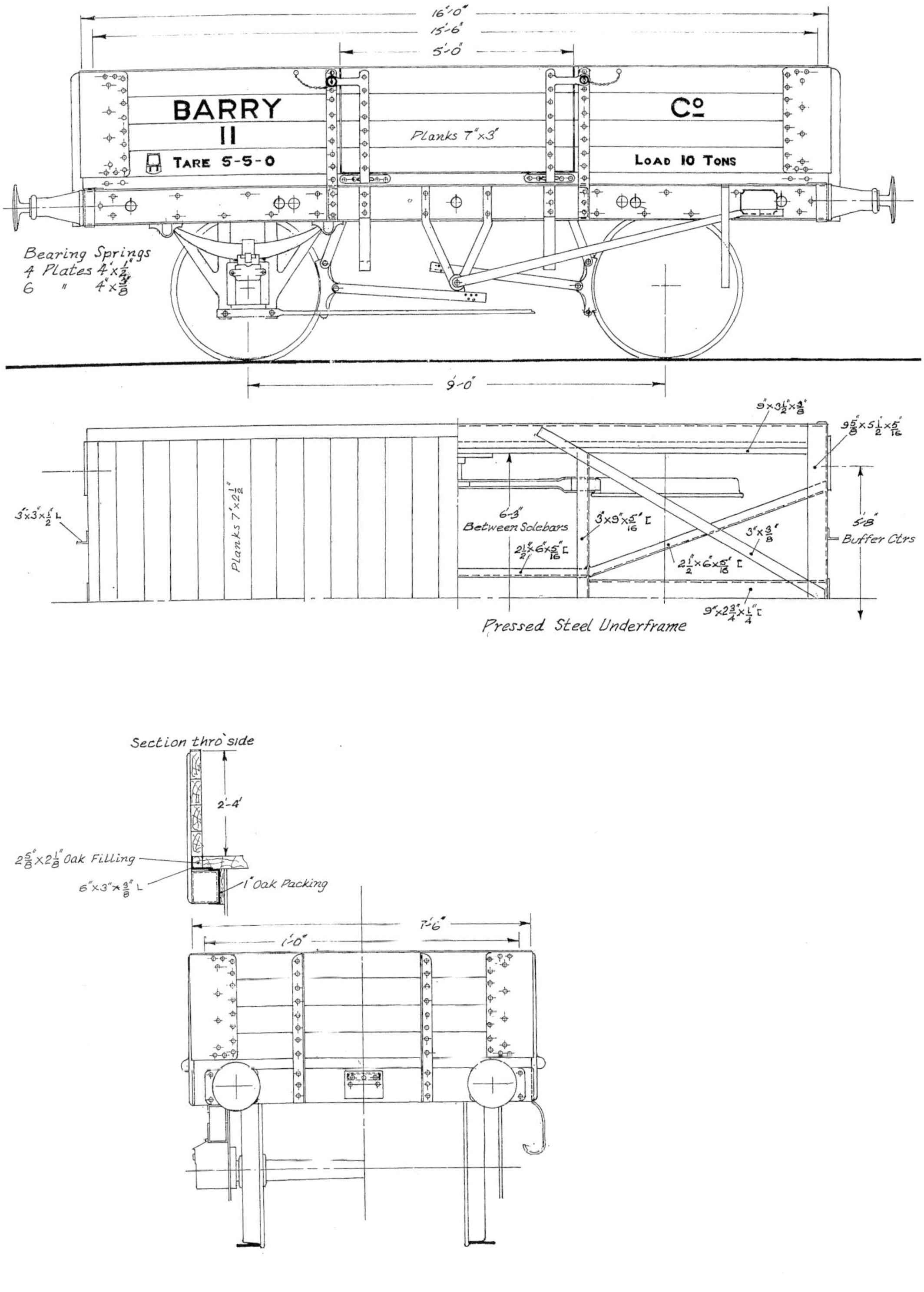

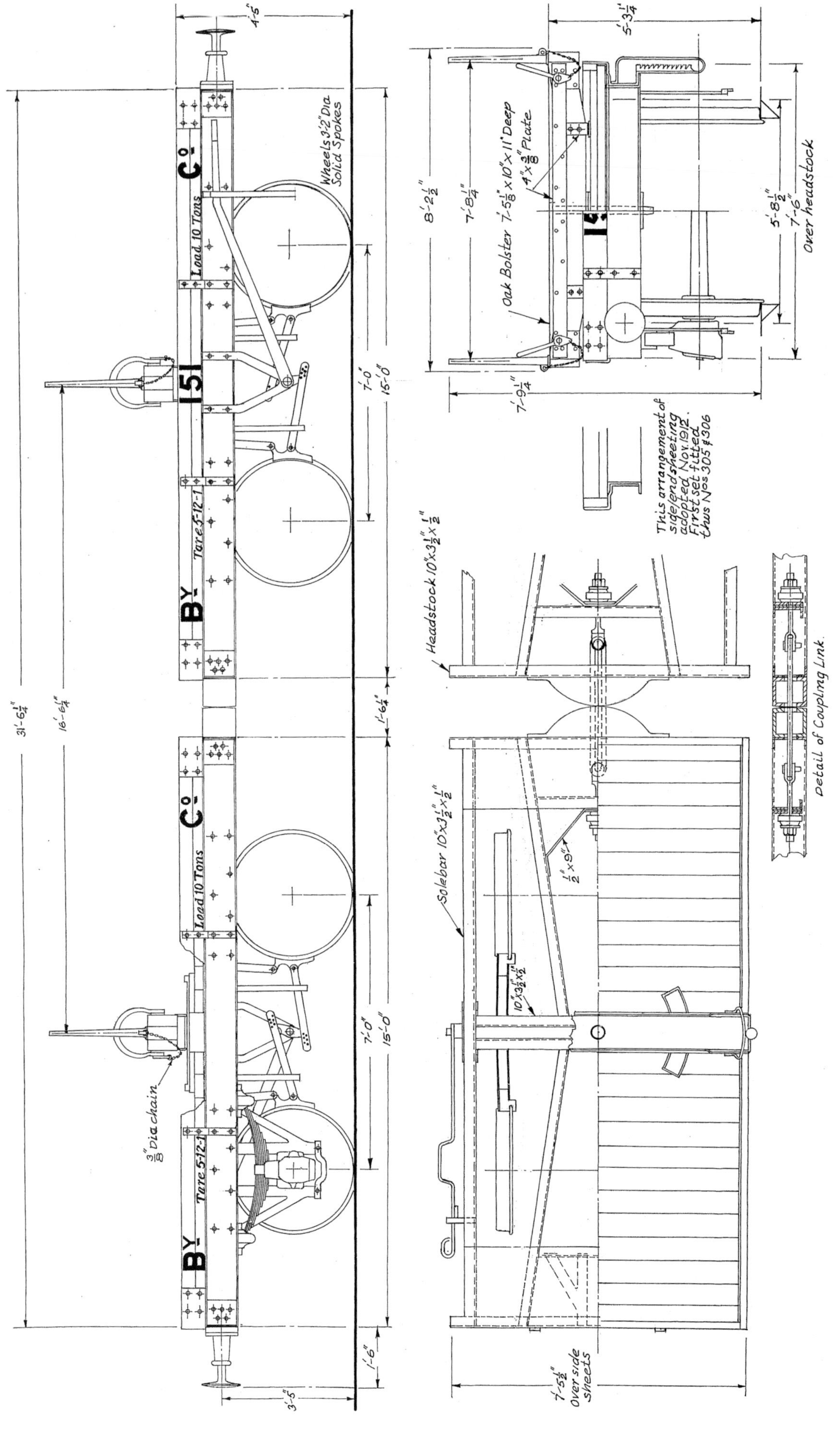

Drawing BRW3 10 Ton Timber Set

Five pairs of timber trucks were built in 1898 by Harrison & Camm, numbered from 293 to 302, and a further ten pairs in 1890/1 by the Swansea Wagon Co, Nos 61–66 and 141–154. The Harrison & Camm wagons were very similar to No. 151. All had been withdrawn by June 1939.

Minutes of the Capital Expenditure Committee
3 June 1897

Tenders received for the supply of 15 pairs of timber trucks were submitted and it was resolved to recommend for acceptance that of Messrs Harrison & Camm for the sum of £2,025.

[However, only five pairs were ordered]

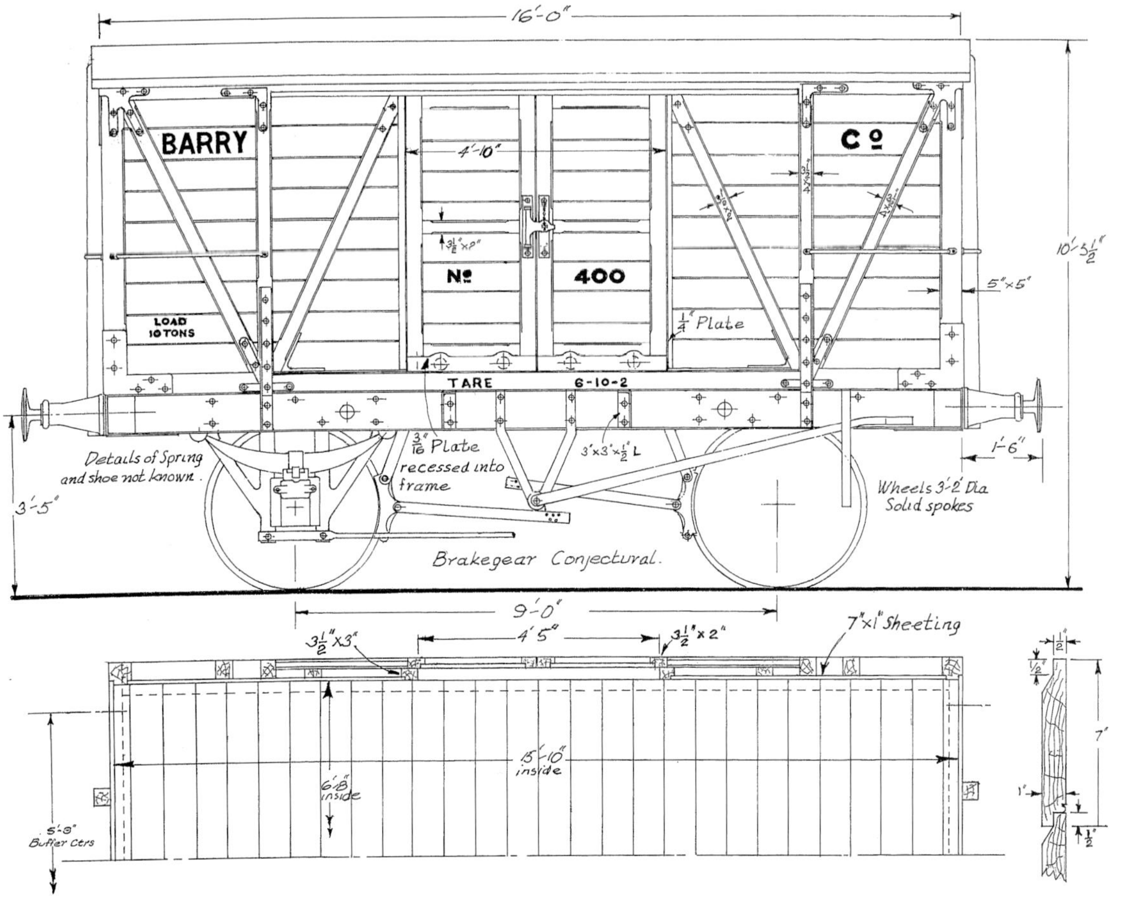

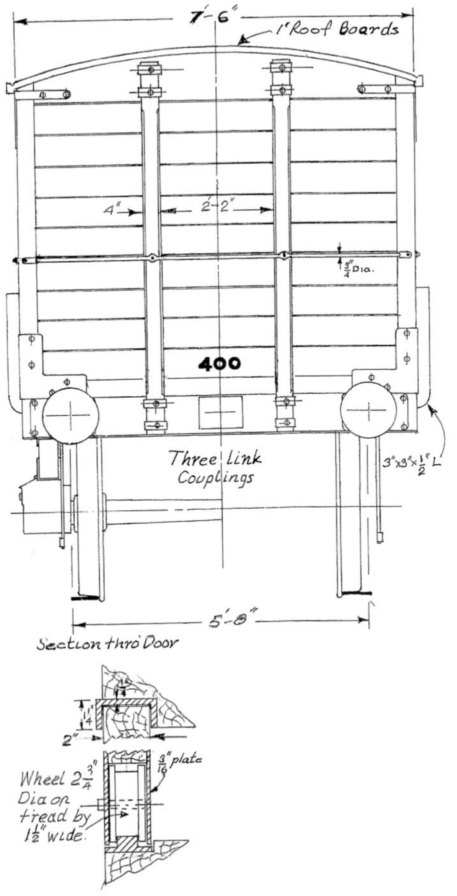

DRAWING BRW4 10 TON COVERED VAN

Just ten vans were built to this design in 1896 by Brown, Marshalls & Co. Apart from a batch of 12 ordered from Gloucester in 1902, of which that in the photograph is an example, all further vans were of the "Iron Mink" variety. No. 400 was withdrawn in August 1928, No. 653 in May 1928.

GLOUCESTER RAILWAY CARRIAGE & WAGON CO. LTD, GLOUCESTERSHIRE ARCHIVES

Drawing BRW5 10 Ton Cattle Truck

Built 1896 by Brown, Marshalls & Co. as one of a pair, costing £232, which sufficed until 1904 when a further batch to an amended design were obtained from Cravens. No. 501 was withdrawn in October 1933.

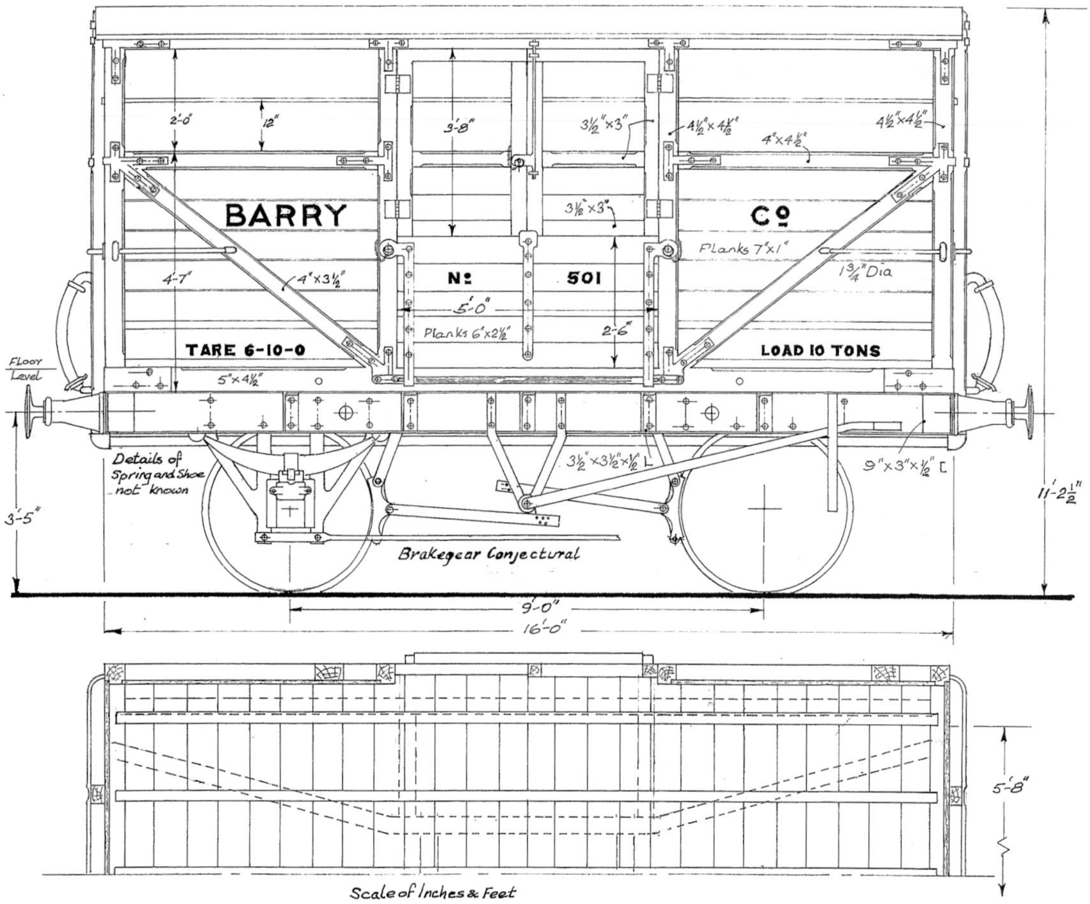

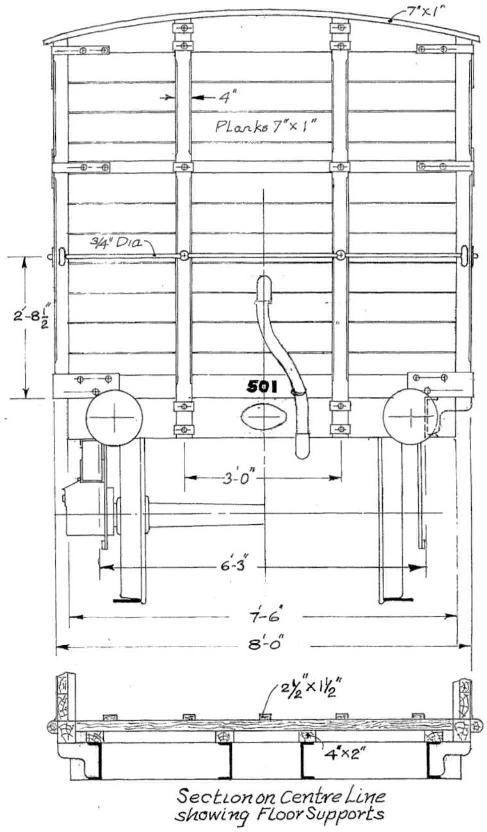

Manager's Report
30 November 1903
Rolling stock – cattle trucks

I beg to draw your attention to the number of cattle trucks owned by the Company. There are only two and this number is totally inadequate to cope with the traffic.

The Great Western Railway have assisted us a great deal and they have not objected to our sending their cattle trucks onto other Company's lines, but of course this procedure is costly from this Company's point of view because we have to pay the wrong sending charges to the extent of 3s per truck per day. The Taff Vale Railway have disappointed us several times lately and they have now informed that they are not able to supply us with cattle wagons as they have not sufficient for their own service.

We are often put to great straits on account of being short of wagons and I consider it very necessary to increase our stock by six.

Will you please consider this matter and instruct me.

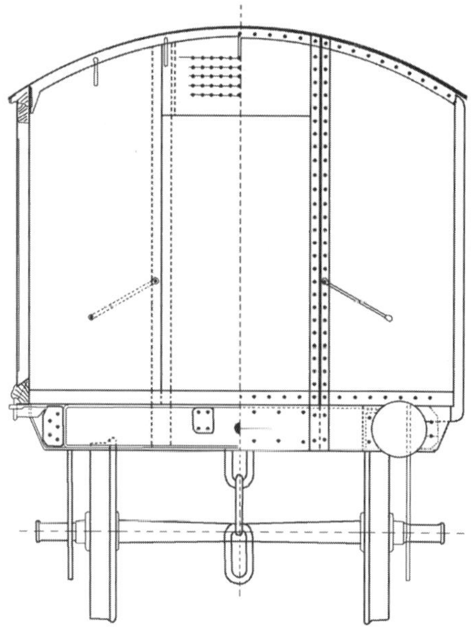

Metropolitan Carriage, Wagon & Finance Co.

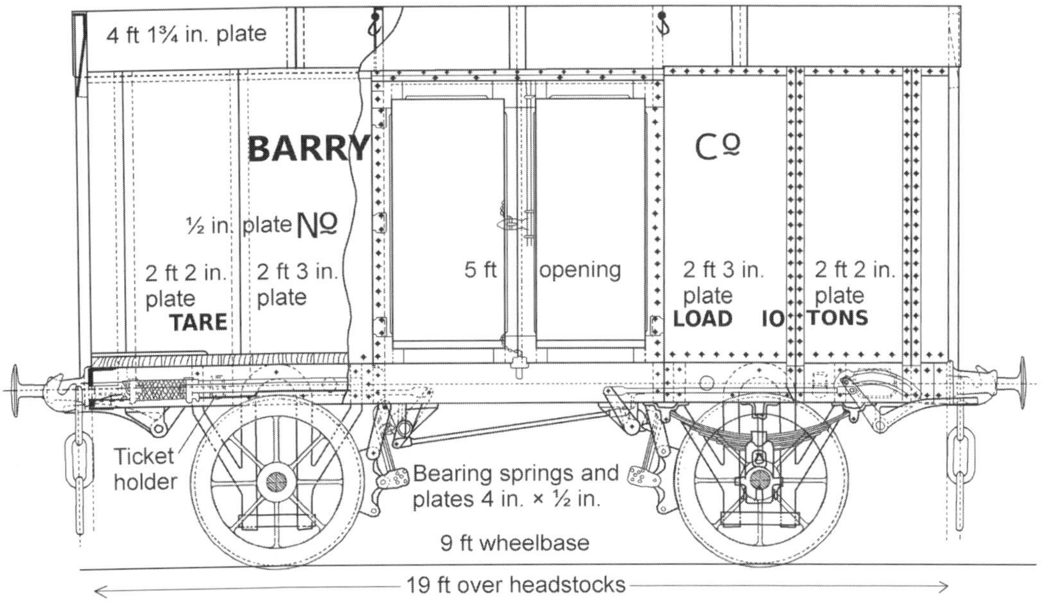

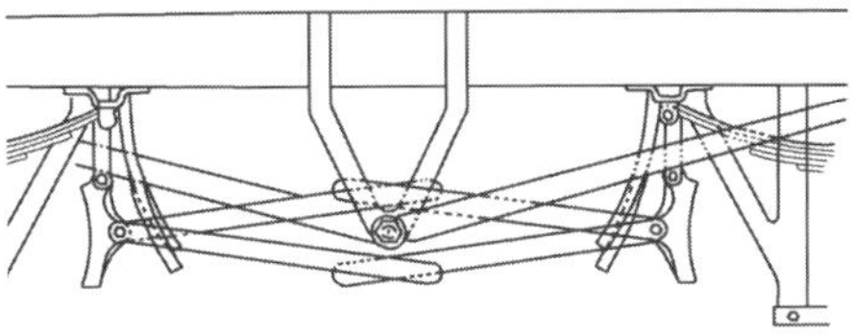

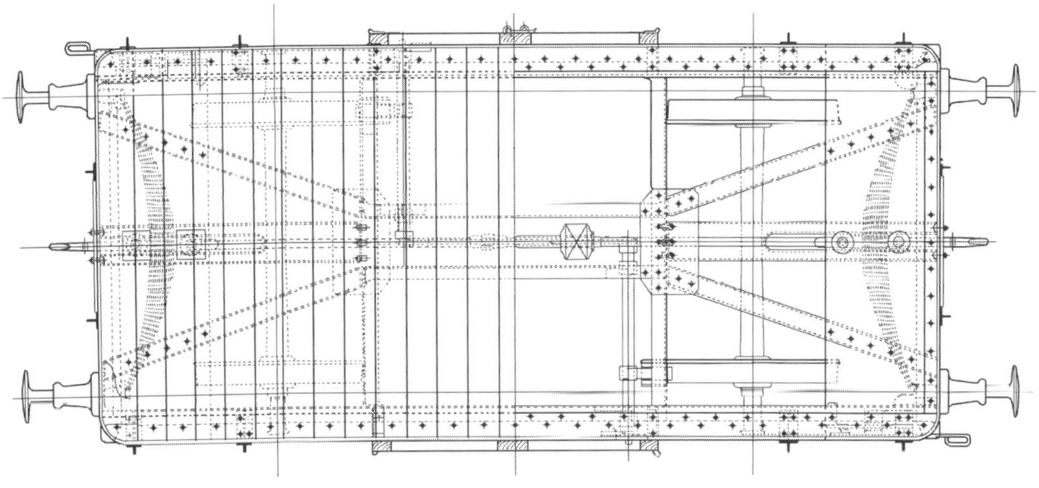

Redrawn by Jonathan David

Drawing BRW6 10 Ton Iron Bodied Van

By 1922 the company owned no less than 730 "Iron Minks", similar in design to those used by several of the companies operating in South Wales. The first batch of 30 was obtained from G R Turner in 1904. The next 150 were also obtained from Turner but later batches were supplied by Midland Carriage & Wagon, S J Claye and Metropolitan. There were minor differences between batches. The first 130 had pressed steel underframes, the remainder steel channel. The last of these vans was withdrawn after nationalisation.

These drawings have been prepared for this book from the official General Arrangement drawing for the first batch of wagons built by Turners. However, the type of brake shown on the GA has not been seen in any photographs, and it is strongly suspected that they were built with conventional brakes as shown below the side elevation and fitted to subsequent orders.

The photograph is of the first of the 1914 Metropolitan built batch and is therefore slightly different from the drawing.

Locomotives

For Notes and codes used in the "Rebuilt" column, see pages 124 and 125.

Barry Railway						Later history			Drawing No.
BR No.	Year built	Wheel arrangement	Class	Builder	Works No.	Renumbered	Rebuilt	Withdrawn	
1	1888	0-6-0T	A	Sharp, Stewart	3449	699	B3A, C, S 5/23	4/31S	BRL1
2	1888	0-6-0T	A	Sharp, Stewart	3450	700	N/A	10/26S	BRL1
3	1888	0-6-0T	A	Sharp, Stewart	3451	702	N/A	4/31L	BRL1
4	1888	0-6-0T	A	Sharp, Stewart	3452	703	N/A	5/32S	BRL1
5	1888	0-6-0T	A	Sharp, Stewart	3453	706	N/A	11/30	BRL1
6	1888	0-6-2T	B	Sharp, Stewart	3454	198	S/9, B, 8/24	1/48	BRL2
7	1888	0-6-2T	B	Sharp, Stewart	3455	199	N/A	8/32L	BRL2
8	1888	0-6-2T	B	Sharp, Stewart	3456	200	N/A	11/45	BRL2
9	1889	0-6-2T	B	Sharp, Stewart	3457	201 Note 6	N/A	10/22	BRL2
10	1889	0-6-2T	B	Sharp, Stewart	3458	203	N/A	7/32L	BRL2
11	1889	0-6-2T	B	Sharp, Stewart	3459	212	N/A	5/32L	BRL2
12	1889	0-6-2T	B	Sharp, Stewart	3460	206	N/A	7/34L	BRL2
13	1889	0-6-2T	B	Sharp, Stewart	3461	205	S/9, 6/24	12/45	BRL2
14	1889	0-6-2T	B	Sharp, Stewart	3462	208	S/9, 8/24	7/32	BRL2
15	1889	0-6-2T	B	Sharp, Stewart	3463	209	S/9, 8/24	5/31	BRL2
16	1889	0-6-2T	B	Sharp, Stewart	3464	210	S/9, 6/24	7/38	BRL2
17	1889	0-6-2T	B	Sharp, Stewart	3465	211	N/A	8/32	BRL2
18	1889	0-6-2T	B	Sharp, Stewart	3466	212	S/9, 6/24	7/48	BRL2
19	1889	0-6-2T	B	Sharp, Stewart	3467	213	N/A	1/49 Note 7	BRL2
20	1889	0-6-2T	B	Sharp, Stewart	3468	214	S/9, 6/29	10/47	BRL2
21	1889	2-4-0T, Note 1	C	Sharp, Stewart	3528	1322	B, C, Metro, 5/24	5/28	BRL5
22	1889	2-4-0T, Note 1	C	Sharp, Stewart	3529	1323	N/A	8/26	BRL5
23	1890	0-6-2T	B	Sharp, Stewart	3571	223	S/9, 6/24	8/32	BRL2
24	1890	0-6-2T	B	Sharp, Stewart	3572	224	N/A	7/47	BRL2
25	1890	0-6-2T	B	Sharp, Stewart	3573	225	N/A	7/32L	BRL2
26	1890	0-6-2T	B	Sharp, Stewart	3574	226	N/A	11/37	BRL2
27	1890	0-6-2T	B	Sharp, Stewart	3575	227	N/A	7/34S	BRL2
28	1890	0-6-2T	B	Sharp, Stewart	3576	228	B3A, 12/23	5/32	BRL2
29	1890	0-6-2T	B	Sharp, Stewart	3577	229	B3A, 12/23	11/34L	BRL2
30	1890	0-6-2T	B	Sharp, Stewart	3578	230	N/A	11/36	BRL2
31	1890	0-6-2T	B	Sharp, Stewart	3579	231	S/9, 8/24	10/49	BRL2
32	1890	0-6-2T	B	Sharp, Stewart	3580	232	S/9, 8/24	8/32	BRL2
33	1889	0-6-0T	E	Hudswell, Clarke	331	781	N/A	7/32S	BRL9
34	1889	0-6-0T	E	Hudswell, Clarke	332	782	C, S, 3/23	6/39S	BRL9
35	Del. 1889	0-8-0	D	Sharp, Stewart	3446	1387	BH, 6/22	11/27L	BRL8, Note 2
36	Del. 1889	0-8-0	D	Sharp, Stewart	3447	1388	C, S, 2/25	8/28	BRL8, Note 2
37	1890	2-4-0T	C	Sharp, Stewart	3610	N/A	N/A	8/1898, Note 3	BRL5
37	1900	0-6-0ST	F	Sharp, Stewart	4593	708	B3B, 3/24	8/32	BRL11
38	1890	0-6-2T	B1	Sharp, Stewart	3598	233	S/9, 6/24	8/32	BRL3
39	1890	0-6-2T	B1	Sharp, Stewart	3599	234	N/A	6/36	BRL3

No. 133/754 – Powell Duffryn Asssociated Collieries Ltd, Bargoed Colliery, scrapped *c.* 07/60 still numbered 754 but rebuilt as a side tank
No. 134/776 – Powell Duffryn Steam Coal Co Ltd, Aberaman Colliery, scrapped 04/60 still numbered 776
No. 138/780 – Burnyeat Brown & Co Ltd, Nine Mile Point Colliery, still at Hafodyrynys Colliery 03/64 still numbered 780; scrapped 6/64
Information in this table is extracted from "The Locomotives of the Great Western Railway. Part Ten. Absorbed engines 1922-1947" (RCTS, 1966). Further information can be found in that volume

MOTOR CARS

As built					As rebuilt		Later history		Drawing No.
BR No.	Year built	Builder (power units)	Builder (coach bodies)	Capacity	BR No.	Capacity	Re-numbered	Withdrawn	
1	1905	North British Locomotive Co.	R Y Pickering & Co.	10 First Class + 40 Third Class + Luggage	177	14 First Class + 56 Third Class	6131	11/51	BRM1, BRM2 after rebuilding
2	1905	North British Locomotive Co.	R Y Pickering & Co.	10 First Class + 40 Third Class + Luggage	178	14 First Class + 56 Third Class (Note 1)	4303 (Note 2)	5/51	BRM1, BRM2 after rebuilding

Notes
1. Converted to a 2nd/3rd Composite before the grouping.
2. Initially allocated the number 6134 but this was never carried. Converted to Third class only with 70 seats and renumbered 4303.

Passenger Rolling Stock

Barry Railway							Later history			Drg No
No.	Date built	Wheels	Type	Compart-ments	Length	Builder	Renum-bered*	Withdrawn	Comments	
4 wheel brake vans nos 1-4 and 5-8 renumbered at an early stage as nos 201-204 and 197-200. Se also page 90.										
1	1888	4	Third	5 × 3rd	27 ft 0 in.	Metropolitan	288*	10/1925		
2	1888	4	Third	5 × 3rd	27 ft 0 in.	Metropolitan	290*	11/1923		
3	1888	4	Third	5 × 3rd	27 ft 0 in.	Metropolitan	80966	1/1931	Note 1	
4	1888	4	Third	5 × 3rd	27 ft 0 in.	Metropolitan	292*	12/1923		
5	1888	4	Third	5 × 3rd	27 ft 0 in.	Metropolitan	294*	10/1925		
6	1888	4	Third	5 × 3rd	27 ft 0 in.	Metropolitan		Pre 12/1921	Note 1	
7	1888	4	Third	5 × 3rd	27 ft 0 in.	Metropolitan	298*	10/1925		
8	1888	4	Third	5 × 3rd	27 ft 0 in.	Metropolitan	299*	12/1923		
9	1888	4	Third	5 × 3rd	27 ft 0 in.	Metropolitan	300*	10/1925		
10	1888	4	Third	5 × 3rd	27 ft 0 in.	Metropolitan	305*	11/1923		
11	1893	6	Third	6 × 3rd	33 ft 10½ in.	Brown, Marshalls & Co	351	12/1926		
12	1893	6	Third	6 × 3rd	33 ft 10½ in.	Brown, Marshalls & Co	352	9/1928		
13	1893	6	Third	6 × 3rd	33 ft 10½ in.	Brown, Marshalls & Co	354	12/1927		
14	1893	6	Third	6 × 3rd	33 ft 10½ in.	Brown, Marshalls & Co	356	11/1927		
15	1893	6	Third	6 × 3rd	33 ft 10½ in.	Brown, Marshalls & Co	363	12/1927	Note 2	
16	1893	6	Third	6 × 3rd	33 ft 10½ in.	Brown, Marshalls & Co	368	12/1925		
17	1889	4	Third	5 × 3rd	27 ft 0 in.	Metropolitan	306*	10/1925		
18	1889	4	Third	5 × 3rd	27 ft 0 in.	Metropolitan	308*	10/1925		
19	1889	4	Third	5 × 3rd	27 ft 0 in.	Metropolitan	315*	10/1923		
20	1889	4	Third	5 × 3rd	27 ft 0 in.	Metropolitan	320*	12/1923		
21	1889	4	Third	5 × 3rd	27 ft 0 in.	Metropolitan	336*	10/1925		
22	1889	4	Third	5 × 3rd	27 ft 0 in.	Metropolitan	80967*	9/1928	Note 1	
23	1889	4	Third	5 × 3rd	27 ft 0 in.	Metropolitan	343*	10/1923		
24	1889	4	Third	5 × 3rd	27 ft 0 in.	Metropolitan	346*	10/1923		
25	1889	4	Third	5 × 3rd	27 ft 0 in.	Metropolitan	348*	10/1925		
26	1889	4	Third	5 × 3rd	27 ft 0 in.	Metropolitan	349*	3/1929		
27	1895	6	Third	6 × 3rd	33 ft 10½ in.	Gloucester	370	11/1927		BRC1
28	1895	6	Third	6 × 3rd	33 ft 10½ in.	Gloucester	371	11/1927		BRC1
29	1895	6	Third	6 × 3rd	33 ft 10½ in.	Gloucester	374	12/1927		BRC1
30	1895	6	Third	6 × 3rd	33 ft 10½ in.	Gloucester	375	12/1927		BRC1
31	1895	6	Third	6 × 3rd	33 ft 10½ in.	Gloucester	376	12/1927		BRC1
32	1895	6	Third	6 × 3rd	33 ft 10½ in.	Gloucester	377	12/1927		BRC1
33	1895	6	Third	6 × 3rd	33 ft 10½ in.	Gloucester	379	11/1927		BRC1
34	1895	6	Third	6 × 3rd	33 ft 10½ in.	Gloucester	380	12/1927		BRC1
35	1895	6	Third	6 × 3rd	33 ft 10½ in.	Gloucester	383*	10/1924		BRC1
36	1895	6	Third	6 × 3rd	33 ft 10½ in.	Gloucester	409	12/1927		BRC1
37	1895	6	Third	6 × 3rd	33 ft 10½ in.	Gloucester	119	12/1927	Note 3	BRC1
38	1895	6	Third	6 × 3rd	33 ft 10½ in.	Gloucester	124	12/1927	Notes 3, 4	BRC1
39	1895	6	Third	6 × 3rd	33 ft 10½ in.	Gloucester	415	12/1927		BRC1
40	1895	6	Third	6 × 3rd	33 ft 10½ in.	Gloucester	417	11/1927		BRC1

Barry Railway							Later history			Drg No.
No.	Date built	Wheels	Type	Compart-ments	Length	Builder	Renum-bered*	Withdrawn	Comments	
41	1895	6	Third	6 × 3rd	33 ft 10½ in.	Gloucester	428	12/1927		BRC1
42	1896	6	Third	6 × 3rd	33 ft 10½ in.	Ashbury	440	12/1927		
43	1896	6	Third	6 × 3rd	33 ft 10½ in.	Ashbury	446	12/1928		
44	1896	6	Third	6 × 3rd	33 ft 10½ in.	Ashbury	450	10/1930		
45	1896	6	Third	6 × 3rd	33 ft 10½ in.	Ashbury	465	12/1928	Note 5	
46	1896	6	Third	6 × 3rd	33 ft 10½ in.	Ashbury	472	12/1927		
47	1896	6	Third	6 × 3rd	33 ft 10½ in.	Ashbury	474	12/1928		
48	1896	6	Third	6 × 3rd	33 ft 10½ in.	Ashbury	484	12/1927		
49	1896	6	Third	6 × 3rd	33 ft 10½ in.	Ashbury	486	12/1928		
50	1896	6	Third	6 × 3rd	33 ft 10½ in.	Ashbury	491	8/1928		
51	1896	6	Third	6 × 3rd	33 ft 10½ in.	Ashbury	494	12/1927		
52	1896	6	Third	6 × 3rd	33 ft 10½ in.	Ashbury	499	12/1928		
53	1896	6	Third	6 × 3rd	33 ft 10½ in.	Ashbury	502	10/1928		
54	1896	6	Third	6 × 3rd	33 ft 10½ in.	Ashbury	512	11/1928		
55	1896	6	Third	6 × 3rd	33 ft 10½ in.	Ashbury	513	11/1929		
56	1896	6	Third	6 × 3rd	33 ft 10½ in.	Ashbury	518	12/1928		
57	1896	6	Third	6 × 3rd	33 ft 10½ in.	Ashbury	524	10/1928		
58	1896	6	Third	6 × 3rd	33 ft 10½ in.	Ashbury	528	1/1929		
59	1896	6	Third	6 × 3rd	33 ft 10½ in.	Ashbury	532	12/1928		
60	1896	6	Third	6 × 3rd	33 ft 10½ in.	Ashbury	534	12/1928		
61	1896	6	Third	6 × 3rd	33 ft 10½ in.	Ashbury	544	12/1928		
62	1896	6	Third	6 × 3rd	33 ft 10½ in.	Ashbury	553	1/1929		
63	1896	6	Third	6 × 3rd	33 ft 10½ in.	Ashbury	557	12/1935		
64	1896	6	Third	6 × 3rd	33 ft 10½ in.	Ashbury	559	6/1928		
65	1896	6	Third	6 × 3rd	33 ft 10½ in.	Ashbury	562	12/1928		
66	1896	6	Third	6 × 3rd	33 ft 10½ in.	Ashbury	575	12/1927		
67	1896	6	Third	6 × 3rd	33 ft 10½ in.	Ashbury	577	12/1927		
68	1896	6	Third	6 × 3rd	33 ft 10½ in.	Ashbury	579	11/1929		
69	1896	6	Third	6 × 3rd	33 ft 10½ in.	Ashbury	585*	11/1924		
70	1896	6	Third	6 × 3rd	33 ft 10½ in.	Ashbury	586	11/1929		
71	1896	6	Third	6 × 3rd	33 ft 10½ in.	Ashbury	589	12/1928	Note 6	
72	1896	6	Third	6 × 3rd	33 ft 10½ in.	Ashbury	593	10/1929		
73	1896	6	Third	6 × 3rd	33 ft 10½ in.	Ashbury	594	11/1928		
74	1896	6	Third	6 × 3rd	33 ft 10½ in.	Ashbury	727	2/1936		
75	1900	6	Third	6 × 3rd	33 ft 10½ in.	Oldbury	760	11/1932		
76	1900	6	Third	6 × 3rd	33 ft 10½ in.	Oldbury	761	11/1932	Note 7	
77	1900	6	Third	6 × 3rd	33 ft 10½ in.	Oldbury	773*	11/1925		
78	1900	6	Third	6 × 3rd	33 ft 10½ in.	Oldbury	775	10/1931		
79	1900	6	Third	6 × 3rd	33 ft 10½ in.	Oldbury	776	3/1936		
80	1900	6	Third	6 × 3rd	33 ft 10½ in.	Oldbury	781	10/1931		
81	1900	6	Third	6 × 3rd	33 ft 10½ in.	Oldbury	784	3/1933		
82	1900	6	Third	6 × 3rd	33 ft 10½ in.	Oldbury	788	11/1932		
83	1900	6	Third	6 × 3rd	33 ft 10½ in.	Oldbury	789	10/1931	Note 7	
84	1900	6	Third	6 × 3rd	33 ft 10½ in.	Oldbury	790	10/1931		
85	1900	6	Third	6 × 3rd	33 ft 10½ in.	Oldbury	799	11/1932		
86	1900	6	Third	6 × 3rd	33 ft 10½ in.	Oldbury	804	12/1935		
87	1900	6	Third	6 × 3rd	33 ft 10½ in.	Oldbury	822	5/1936		

Barry Railway							Later history			Drg No.
No.	Date built	Wheels	Type	Compart-ments	Length	Builder	Renum-bered*	Withdrawn	Comments	
88	1900	6	Third	6 × 3rd	33 ft 10½ in.	Oldbury	823	10/1931	Note 7	
89	1900	6	Third	6 × 3rd	33 ft 10½ in.	Oldbury	824	10/1931	Note 7	
90		6	Third	5 × 3rd	31 ft 11 in.	L&YR	825	11/1929	Acquired 1916	
91		6	Third	5 × 3rd	31 ft 11 in.	L&YR	826	11/1929	Acquired 1916	
92		6	Third	5 × 3rd	31 ft 11 in.	L&YR	827	10/1930	Acquired 1916	
93		6	Third	5 × 3rd	31 ft 11 in.	L&YR	828	11/1929	Acquired 1916	
94		6	Third	5 × 3rd	31 ft 11 in.	L&YR	829	10/1928	Acquired 1916	
95		6	Third	5 × 3rd	31 ft 11 in.	L&YR	830	11/1929	Acquired 1916	
96		6	Third	5 × 3rd	31 ft 11 in.	L&YR	831*	11/1925	Acquired 1916	
97		6	Third	5 × 3rd	31 ft 11 in.	L&YR	832	11/1929	Acquired 1916	
98		6	Third	5 × 3rd	31 ft 11 in.	L&YR	833*	11/1925	Acquired 1916	
99		6	Third	5 × 3rd	31 ft 11 in.	L&YR	834	10/1928	Acquired 1916	
100		6	Third	5 × 3rd	31 ft 11 in.	L&YR	835	12/1927	Acquired 1916	
101		6	Third	5 × 3rd	31 ft 11 in.	L&YR	836*	11/1925	Acquired 1916	
102		6	Third	5 × 3rd	32 ft 11 in.	L&YR	837	10/1930	Acquired 1916	
103		6	Third	5 × 3rd	31 ft 11 in.	L&YR	838	8/1930	Acquired 1916	
104		6	Third	5 × 3rd	31 ft 11 in.	L&YR	839*	11/1925	Acquired 1916	
105		6	Third	5 × 3rd	31 ft 11 in.	L&YR	841	10/1928	Acquired 1916	
150	1888	4	Compo	2 × 1st, 2 × 2nd	27 ft 0 in.	Metropolitan	6026*	11/1923		
151	1888	4	Compo	2 × 1st, 2 × 2nd	27 ft 0 in.	Metropolitan	6030*	11/1923		
152	1888	4	Compo	2 × 1st, 2 × 2nd	27 ft 0 in.	Metropolitan	6035*	10/1923	Auto-fitted and close coupled with 38	
153	1888	4	Compo	2 × 1st, 2 × 2nd	27 ft 0 in.	Metropolitan	6016*	10/1925		
154	1888	4	Compo	2 × 1st, 2 × 2nd	27 ft 0 in.	Metropolitan	6020*	10/1925		
155	1888	4	Compo	2 × 1st, 2 × 2nd	27 ft 0 in.	Metropolitan	4307	10/1925		
156	1888	4	Compo	2 × 1st, 2 × 2nd	27 ft 0 in.	Metropolitan	6037*	11/1923		

Barry Railway							Later history			Drg No.
No.	Date built	Wheels	Type	Compart-ments	Length	Builder	Renum-bered*	Withdrawn	Comments	
157	1888	4	Compo	2 × 1st, 2 × 2nd	27 ft 0 in.	Metropolitan	6024*	1/1924	Auto-fitted and close coupled with 37	
158	1895	6	Compo	2 × 1st, 3 × 2nd	33 ft 11 in.	Ashbury	6042	11/1926		BRC2
159	1895	6	Compo	2 × 1st, 3 × 2nd	33 ft 11 in.	Ashbury	6044	3/1926		BRC2
160	1895	6	Compo	2 × 1st, 3 × 2nd	33 ft 11 in.	Ashbury	6046	12/1927		BRC2
161	1895	6	Compo	2 × 1st, 3 × 2nd	33 ft 11 in.	Ashbury	6048	12/1927		BRC2
162	1895	6	Compo	2 × 1st, 3 × 2nd	33 ft 11 in.	Ashbury	6052	12/1927		BRC2
163	1895	6	Compo	2 × 1st, 3 × 2nd	33 ft 11 in.	Ashbury	6058	11/1928		BRC2
164	1896	6	Compo	2 × 1st, 3 × 2nd	33 ft 11 in.	Ashbury	6060	11/1928		BRC2
165	1896	6	Compo	2 × 1st, 3 × 2nd	33 ft 11 in.	Ashbury	6061	11/1929		BRC2
166	1896	6	Compo	2 × 1st, 3 × 2nd	33 ft 11 in.	Ashbury	6062	12/1932		BRC2
167	1896	6	Compo	2 × 1st, 3 × 2nd	33 ft 11 in.	Ashbury	6063	12/1928		BRC2
168	1898	6	Compo	2 × 1st, 3 × 2nd	33 ft 11 in.	Ashbury	6064	12/1928		BRC2
169	1898	6	Compo	2 × 1st, 3 × 2nd	33 ft 11 in.	Ashbury	6069	12/1928		BRC2
170	1898	6	Compo	2 × 1st, 3 × 2nd	33 ft 11 in.	Ashbury	6071	8/1928		BRC2
171	1898	6	Compo	2 × 1st, 3 × 2nd	33 ft 11 in.	Ashbury	6084	2/1936		BRC2
172	1898	6	Compo	2 × 1st, 3 × 2nd	33 ft 11 in.	Ashbury	6086	12/1928		BRC2
173	1898	6	Compo	2 × 1st, 3 × 2nd	33 ft 11 in.	Ashbury	6095	12/1928	Note 8	BRC2
174	1898	6	Compo	2 × 1st, 3 × 2nd	33 ft 11 in.	Ashbury	6099*	11/1924		BRC2
175	1900	6	Compo	2 × 1st, 3 × 2nd	33 ft 11 in.	Oldbury	6106	3/1936		BRC2
176	1900	6	Compo	2 × 1st, 3 × 2nd	33 ft 11 in.	Oldbury	6109	2/1932		BRC2
177	1905	8	Bk Compo	2 × 1st, 2 × 3rd, 1 × comb. Gd & Lug	60 ft 10 in.	Pickering	6131	5/1951	Formerly steam Motor Car	
178	1905	8	Bk Compo	2 × 1st, 2 × 3rd, 1 × comb. Gd & Lug	60 ft 10 in.	Pickering	4303	11/1951	Formerly steam Motor Car, see text	
179		6	Compo	2 × 2nd, 3 × 3rd	32 ft 11 in.	L&YR	6113	8/1930	Acquired 1916	

Barry Railway							Later history			Drg No.
No.	Date built	Wheels	Type	Compart-ments	Length	Builder	Renum-bered*	Withdrawn	Comments	
180		6	Compo	2 × 1st, 3 × 2nd	32 ft 11 in.	L&YR	6115	8/1930	Acquired 1916	
181		6	Compo	2 × 1st, 3 × 2nd	32 ft 11 in.	L&YR	6117	11/1929	Acquired 1916	
182		6	Compo	2 × 1st, 3 × 2nd	32 ft 11 in.	L&YR	6120	9/1928	Acquired 1916	
183		6	Compo	2 × 1st, 3 × 2nd	32 ft 11 in.	L&YR	6122	10/1930	Acquired 1916	
184		6	Compo	2 × 1st, 3 × 2nd	32 ft 11 in.	L&YR	6123	10/1930	Acquired 1916	
197	1888	4	Bk 3rd	1 × 3rd, 1 × Gd, 1 × Lug	21 ft 6 in.	Metropolitan	112*	4/1923		
198	1888	4	Bk Van	1 × Gd, 1 × Lug	21 ft 6 in.	Metropolitan	35*	11/1923		
199	1888	4	Bk Van	1 × Gd, 1 × Lug	21 ft 6 in.	Metropolitan	36*	11/1923		
200	1888	4	Bk Van	1 × Gd, 1 × Lug	21 ft 6 in.	Metropolitan	37*	11/1923		
201	1890	4	Bk Van	1 × Gd, 1 × Lug	21 ft 6 in.	Metropolitan	9*	10/1925		BRC3
202	1890	4	Bk Van	1 × Gd, 1 × Lug	21 ft 6 in.	Metropolitan	24*	10/1925		BRC3
203	1890	4	Bk Van	1 × Gd, 1 × Lug	21 ft 6 in.	Metropolitan	27*	10/1923		BRC3
204	1890	4	Bk Van	1 × Gd, 1 × Lug	21 ft 6 in.	Metropolitan	30*	12/1923		BRC3
205	1895	6	Bk 3rd	2 × 3rd, 1 × Gd, 2 × Lug	33 ft 11 in.	Gloucester	126	3/1926	Note 9	BRC4
206	1895	6	Bk 3rd	2 × 3rd, 1 × Gd, 2 × Lug	33 ft 11 in.	Gloucester	156	12/1927	Note 9	BRC4
207	1895	6	Bk 3rd	2 × 3rd, 1 × Gd, 2 × Lug	33 ft 11 in.	Gloucester	179	10/1926	Note 9	BRC4
208	1895	6	Bk 3rd	2 × 3rd, 1 × Gd, 1 × Lug	33 ft 11 in.	Gloucester	181	4/1928	Note 9	BRC4
209	1896	6	Bk 3rd	2 × 3rd, 1 × Gd, 2 × Lug	33 ft 11 in.	Ashbury	184	11/1929	Note 9	
210	1896	6	Bk 3rd	2 × 3rd, 1 × Gd, 2 × Lug	33 ft 11 in.	Ashbury	185	12/1927	Note 9	
211	1896	6	Bk 3rd	2 × 3rd, 1 × Gd, 2 × Lug	33 ft 11 in.	Ashbury	187	10/1931	Note 9	
212	1896	6	Bk 3rd	2 × 3rd, 1 × Gd, 2 × Lug	33 ft 11 in.	Ashbury	189	12/1928	Note 9	
213	1897	6	Bk 3rd	2 × 3rd, 1 × Gd, 2 × Lug	33 ft 11 in.	Ashbury	194	12/1927	Note 9	
214	1897	6	Bk 3rd	2 × 3rd, 1 × Gd, 2 × Lug	33 ft 11 in.	Ashbury	199	3/1933	Note 9	
215	1897	6	Bk 3rd	2 × 3rd, 1 × Gd, 2 × Lug	33 ft 11 in.	Ashbury	200	1/1929	Note 9	
216	1897	6	Bk 3rd	2 × 3rd, 1 × Gd, 2 x Lug	33 ft 11 in.	Ashbury	38	12/1927	Note 9	

Barry Railway							Later history			Drg No.
No.	Date built	Wheels	Type	Compart- ments	Length	Builder	Renum- bered*	Withdrawn	Comments	
217	1898	6	Bk Van	2 × Gd, 2 × Lug	33 ft 11 in.	Ashbury	39	11/1929	Note 9	
218	1898	6	Bk 3rd	2 × 3rd, 1 × Gd, 2 × Lug	33 ft 11 in.	Ashbury	201	10/1929	Note 9	
219	1898	6	Bk Van	2 × Gd, 2 × Lug	33 ft 11 in.	Ashbury	40	12/1927	Note 9	
220	1898	6	Bk Van	2 × Gd, 2 × Lug	33 ft 11 in.	Ashbury	41	12/1927	Note 9	
221	1900	6	Bk 3rd	3 × 3rd, 1 × Gd, 1 x Lug	33 ft 11 in.	Oldbury	207	10/1931		
222	1900	6	Bk 3rd	3 × 3rd, 1 × Gd, 1 x Lug	33 ft 11 in.	Oldbury	208	3/1936		
223	1900	6	Bk 3rd	3 × 3rd, 1 × Gd, 1 x Lug	33 ft 11 in.	Oldbury	232	9/1928		
224	1900	6	Bk 3rd	3 × 3rd, 1 × Gd, 1 x Lug	33 ft 11 in.	Oldbury	233	11/1932		
225	1900	6	Bk 3rd	3 × 3rd, 1 × Gd, 1 x Lug	33 ft 11 in.	Oldbury	241	11/1932		
226	1901	6	Bk Van	1 × Gd, 2 × Lug	33 ft 11 in.	Oldbury	42	10/1933		
227		6	Bk 3rd	3 × 3rd, 1 comb. Gd & Lug	31 ft 11 in.	L&YR	242	12/1927	Acquired 1916	
228		6	Bk 3rd	3 × 3rd, 1 comb. Gd & Lug	31 ft 11 in.	L&YR	243*	4/1925	Acquired 1916	
229		6	Bk 3rd	3 × 3rd, 1 comb. Gd & Lug	31 ft 11 in.	L&YR	244	9/1928	Acquired 1916	
230		6	Bk 3rd	3 × 3rd, 1 comb. Gd & Lug	31 ft 11 in.	L&YR	245*	10/1924	Acquired 1916	
231		6	Bk 3rd	3 × 3rd, 1 comb. Gd & Lug	31 ft 11 in.	L&YR	246	11/1929	Acquired 1916	
212		6	Bk 3rd	3 × 3rd, 1 comb. Gd & Lug	31 ft 11 in.	L&YR	247	1/1929	Acquired 1916	
233		6	Bk 3rd	3 × 3rd, 1 comb. Gd & Lug	31 ft 11 in.	L&YR	252	1/1929	Acquired 1916	
234		6	Bk 3rd	3 × 3rd, 1 comb. Gd & Lug	31 ft 11 in.	L&YR	256	9/1928	Acquired 1916	
247	1896	4	Horesbox, 3 stalls + groom		18 ft 0 in.	Ashbury	3*	2/1924		
248	1896	4	Horesbox, 3 stalls + groom		18 ft 0 in.	Ashbury	4	12/1924		
251	1896	4	Carriage truck		18 ft 0 in.	Ashbury	4*	12/1927		
253		4	Carriage Van		18 ft 0 in.	Ashbury	1*	2/1924		

Barry Railway							Later history			Drg No.
No.	Date built	Wheels	Type	Compart-ments	Length	Builder	Renum-bered*	Withdrawn	Comments	
Second series										
2	1920	8	Third	9 × 3rd	51 ft 0 in.	Birmingham	279	3/1958		BRC5
3	1920	8	Third	9 × 3rd	51 ft 0 in.	Birmingham	281	12/1956		BRC5
4	1920	8	Third	9 × 3rd	51 ft 0 in.	Birmingham	282	7/1957		BRC5
6	1920	8	Third	9 × 3rd	51 ft 0 in.	Birmingham	283	1/1959		BRC5
8	1920	8	Third	9 × 3rd	51 ft 0 in.	Birmingham	284	12/1956		BRC5
10	1920	8	Third	9 × 3rd	51 ft 0 in.	Birmingham	287	3/1948		BRC5
150	1920	8	Tri-Compo	3 × 1st, 4 × 2nd, 1 × 3rd	52 ft 6 in.	Birmingham	6125	6/1957		BRC6
151	1920	8	Tri-Compo	3 × 1st, 4 × 2nd, 1 × 3rd	52 ft 6 in.	Birmingham	6126	4/1959		BRC6
152	1920	8	Tri-Compo	3 × 1st, 4 × 2nd, 1 × 3rd	52 ft 6 in.	Birmingham	6127	12/1956		BRC6
156	1920	8	Tri-Compo	3 × 1st, 4 × 2nd, 1 × 3rd	52 ft 6 in.	Birmingham	6130	11/1957		BRC6
197	1920	8	Bk 3rd	5 × 3rd, 1 × G, 1 × Lug	51 ft 0 in.	Birmingham	263	10/1962		BRC7
198	1920	8	Bk 3rd	5 × 3rd, 1 × G, 1 × Lug	51 ft 0 in.	Birmingham	268	10/1962		BRC7
199	1920	8	Bk 3rd	5 × 3rd, 1 × G, 1 × Lug	51 ft 0 in.	Birmingham	274	1/1959		BRC7
200	1920	8	Bk 3rd	5 × 3rd, 1 × G, 1 × Lug	51 ft 0 in.	Birmingham	275	11/1948		BRC7

Notes

* Number allocated but not carried

1. Transferred to Engineers' Department use
2. Under restoration at Hampton Loade, Severn Valley Railway, using four-wheeled CCT underframe. Owned by Barry Railway Carriage Trust
3. In 1909, Thirds Nos. 37 and 38 had one end compartment converted to a Driving-cum-Guard's compartment for auto train working
4. Body only preserved, Cornwall, not restored
5. Body only preserved, not restored. Owned by Barry Railway Carriage Trust
6. Body only preserved, Hertfordshire, not restored
7. One third class compartment upgraded to second class
8. Preserved, at Pontypool & Blaenafon Railway, being restored; using four-wheeled CCT underframe
9. Later most had one or both compartments removed. See page 93. Nos. 208 and 210 are listed by Mountford as having three Third class compartments at the Grouping.

Mineral Brake Vans

See drawings BRB1 and BRB2

Nos.	Year built	Description	Builder	Renumbered	Comments
1-15	1889	Four-wheeled	Metropolitan Carriage & Wagon Co	17969/70/75/80/81/83/89/96/ 18000/22347/49/50/52/22789	No. 9 not taken into GWR stock. All withdrawn by 3/1927
16-25	1890	Four-wheeled	Swansea Wagon Co	22733/34/35/36/37/39/41 22742/43/45	All withdrawn 1926 except Nos. 19 (2/1928) and 21 (10/1929)
26-30	1890	Four-wheeled	Metropolitan Carriage & Wagon Co	22746/31293/35725/58 35886	All withdrawn by 6/1927
32-46	1896	Four-wheeled	Midland Carriage & Wagon Co	35888/89/90/91/92/35906/12/ 35913/16/18/25/26/27/28/33	All withdrawn by 3/1927 except No. 45 (12/1931)
47-61	1900	Four-wheeled	Ashbury Carriage & Wagon Co	35934/38/39/41/42/44/45 35948/50/54/60/61/62/64 35966	All withdrawn by 6/1927
62-73	1915	Six-wheeled	Midland Carriage & Wagon Co	35971/73/77/81/82/83/85 35987/93/98/56481/82	Withdrawal dates not recorded

Goods Traffic Rolling Stock at the Grouping – Revenue Stock

Company abbreviations used throughout these lists:

Ashbury	Ashbury Railway Carriage & Iron Co Ltd, Manchester; later amalgamated with Brown. Marshalls & Co Ltd
Barry Co	Built by Barry Railway Co workshops
Brown, Marshalls	Brown, Marshalls & Co Ltd, Saltley, Birmingham; later amalgamated with the Metropolitan Railway Carriage, Wagon & Finance Co
Gloucester	Gloucester Railway Carriage & Wagon Company Ltd, Gloucester
Harrison & Camm	Harrison & Camm; Rotherham and Sheffield
S J Claye	S J Claye Ltd; Long Eaton, Nottingham
Metropolitan	Metropolitan Railway Carriage, Wagon & Finance Co Ltd, Saltley, Birmingham; later amalgamated with others to form Metropolitan Cammell Carriage, Wagon and Finance Company Ltd, commonly known as Metro-Cammell
Midland	Midland Railway Carriage & Wagon Co, Birmingham; eventually became part of the Metropolitan Cammell Carriage, Wagon & Finance Company Ltd
Swansea	Swansea Carriage & Wagon Co Ltd, Swansea; not to be confused with the later Swansea Wagon Co Ltd
Turner	G R Turner, Langley Mill

For Notes, see page 139.

Barry Railway history				Subsequent history		Notes
BR No.	Built	Builder	Description	Renumbered	Withdrawn	
1, 12, 14, 21-49, 52-60, 93-97, 99, 105-07	1912	S J Claye	10 T open, 16 ft over headstocks, 9 ft wheelbase	8989, 9000, 10005, 10283, 10302, 10310, 10311, 10320, 10449, 10531, 10556, 10564, 10593, 10605, 10633, 10643, 10658, 10671, 10689, 10696, 10712, 10720, 10721, 10725, 10743, 10772, 10827, 10842, 10877, 11454, 11455, 11458, 11464, 11466, 11467, 11469, 11471, 11473, 11474, 11477, 11478, 11516, 11518, 11520-22, 11527, 11535, 11538, 11539	Between 3/25 and 1/39	Dwg. BRW1 Note 1
2, 5, 11, 20, 50, 51, 155, 158, 159, 207	1903	Barry Co	10 T open, 16 ft over headstocks, 9 ft wheelbase	8990, 8993, 8999, 10228, 11459, 11460, 11600, 11609, 11610, 11705	Between 2/25 and 1/37, 158 uncertain	Dwg. BRW2
3	1897	Ashbury	10 T open, 16 ft over headstocks, 9 ft wheelbase	8991	4/35	
4, 122, 125, 134, 191, 196, 218	1902	Barry Co	10 T open, 16 ft over headstocks, 9 ft wheelbase	8992, 11564, 11570, 11586, 11675, 11684, 11723	Between 12/28 and 5/33	Dwg BRW2 Note 2
6-8, 15, 19, 87-89, 120, 127, 130, 162, 163, 182, 184, 186, 195	1901	Barry Co	10 T open, 16 ft over headstocks, 9 ft wheelbase	8994-96, 10015, 10201, 11506, 11508, 11510, 11562, 11572, 11582, 11613, 11614, 11659, 11664, 11666, 11682	Between 3/25 and 11/37	Dwg BRW2 Note 2
9, 10, 17, 18, 116, 176, 212	1905	Barry Co	10 T open, 16 ft over headstocks, 9 ft wheelbase	8997, 8998, 10165, 10195, 11557, 11647, 11716	Between 10/26 and 10/32	Dwg BRW2 Note 3
13, 124, 165, 167, 174, 197, 198, 204, 213, 627, 628, 631	1910	Barry Co	10 T open, 16 ft over headstocks, 9 ft wheelbase	10002, 11569, 11619, 11628, 11643, 11685, 11693, 11702, 11717, 22533, 22534, 22537	Between 9/27 and 9/40; 174 and 628 NR, possibly after 1947	
16, 156, 160, 161, 164, 169-72, 188, 189, 192, 194, 209	1904	Barry Co	10 T open, 16 ft over headstocks, 9 ft wheelbase	10025, 11602, 11611, 11612, 11615, 11634, 11535, 11637, 11640, 11671, 11673, 11677, 11681, 11709	Between 12/23 and 12/38; 156 NR, possibly after 1947	Dwg. BRW2

Barry Railway history				Subsequent history		Notes
BR No.	Built	Builder	Description	Renumbered	Withdrawn	
61-66, 141-154	1890/1	Swansea	10 T timber sets, 15 ft, 7 ft wheelbase	8441, 8443-47, 8451, 8453-47, 8451, 8452, 8456, 8458, 8470-75, 8477-79, 8486	Between 5/32 and 3/38	Dwg. BRW3
67-76, 100, 136, 138, 140, 157, 166, 168, 177, 178, 190, 193, 210, 211, 239, 241, 243-50, 252-56, 258, 259, 262-64, 266, 268-76, 278, 279, 281-83, 285-88, 348, 349, 365, 366, 382-84, 387, 397, 411, 414, 423, 427, 433, 443, 446-448, 450, 454, 462, 464, 468, 487, 490, 494, 520, 527, 535, 542, 545, 550, 560, 566, 577, 578, 584, 587, 589-626, 629, 630, 632-641	1920	S J Claye	10 T open, 16 ft over headstocks, 9 ft wheelbase	11479, 11476, 11482, 11483, 11485, 11486, 11488-91, 11528, 11589, 11590, 11599, 11608, 11622, 11631, 11650, 11653, 11674, 11678, 11712, 11713, 11725, 11729, 11732, 11736-39, 11746, 11747, 11749, 11752-57, 11762, 11766, 11772, 11773, 11775, 11779, 11786-89, 11791-93, 11795, 11798, 11801, 11802, 11807, 11809, 11810, 11812, 11814, 11815, 11818, 12212, 12243, 12869, 12932, 15073, 15081, 15083, 15124, 15217, 15255, 15306, 15374, 15401, 15450, 15647, 15696, 15706, 15720, 15741, 15783, 15848, 15879, 15924, 17491, 17494, 17498, 18563, 19256, 19320, 19927, 20063, 20148, 20214, 20258, 20451, 20452, 20480, 20491, 21319, 21337, 21413, 21579, 21650, 21660, 21819, 21849, 21896, 21908, 21953, 21989, 21991, 21995, 21996-98, 22355, 22357, 22374, 22404, 22411, 22415, 22420, 22421, 22426, 22430, 22441, 22445, 22449, 22524-26, 22528-32, 22535, 22536, 22538, 22794, 22798, 22864, 24006, 24086, 24126, 24146, 24158, 24192	Most between **8/25** and 8/37; 42, 520 NR, possible after 1947	Dwg. BRW1; note that 19927 may be an error for 19937
77-86	1890	Swansea	One plank open wagons, 16ft long, 9 ft wheelbase	11494, 11496-11501, 11503-05	Between 12/30 and 9/40	
90, 91, 101, 110, 131, 133	1901	Barry Co	10 T open, 16 ft over headstocks, 9 ft wheelbase	11511, 11513, 11529, 11545, 11583, 11585	Between 10/26 and 8/34	Dwg BRW2 Note 2
92, 181, 183, 185, 187, 208, 412, 588	1903	Barry Co	10 T open, 16 ft over headstocks, 9 ft wheelbase	11514, 11658, 11662, 11665, 11669, 11708, 15260, 21001	Between 1/26 and 12/41	Dwg. BRW2 Note 2
98, 114	1890	Ashbury	10 T open, 16 ft long, 9 ft wheelbase	11525, 11554	9/37, 12/31	
102, 132	1898	Ashbury	10 T open, 16 ft long, 9 ft wheelbase	11531, 11584	8/28, 12/30	
103, 129	1903	Barry Co	10 T open, 15 ft long, 9 ft wheelbase	11532, 11580	7/32, 8/26	
104	1899	Barry Co	10 T open, 16 ft over headstocks, 9 ft wheelbase	11533	11/28	Dwg BRW2 Note 2

Barry Railway history				Subsequent history		Notes
BR No.	Built	Builder	Description	Renumbered	Withdrawn	
108, 111-113, 115, 117-19, 123, 135, 137, 139, 173, 175, 179, 199-201, 205, 214, 215, 217, 240, 242, 251, 257, 260, 261, 265, 267, 277, 280, 284, 354, 362, 369, 374, 391, 392, 434, 435, 463, 467, 480, 488, 492, 522, 533, 568, 574, 1093-1142	1913	S J Claye	10 T open 16 ft over headstocks, 9 ft wheelbase	11542, 11546, 11548, 11550, 11556, 11558, 11560, 11561, 11566, 11588, 11540, 11596, 11642, 11644, 11656, 11695-97, 11703, 11718, 11720, 11722, 11727, 11730, 11751, 11760, 11768, 11771, 11777, 11782, 11800, 11803, 11811, 12355, 12831, 13274, 13538, 15158, 15193, 15467, 15511, 15849, 15902, 17484, 17492, 17496, 18845, 19317, 20278, 20387, 25477, 25483, 25487, 25491, 25495, 25503, 25509, 25516, 25524, 25532, 25535, 25536, 25546, 25543, 25558, 25561, 25572-75, 25577, 25581, 25607, 25610, 25613, 25614, 25618, 25621, 25622, 25631, 25637, 25643, 25648, 25654, 25656, 25667, 25670, 25674, 25675, 25679, 25682-84, 25687, 25692, 25694, 25700, 25702, 25704, 25705	Between 4/23 and 7/39	Dwg. BRW2 Note 1
109, 121	?	?	10 T open 15 ft over headstocks, 9 ft wheelbase	11544, 11563	8/31, 10/28	
126, 128, 202, 203	?	?	10 T open, 16 ft over headstocks, 9 ft wheelbase	11571, 11574, 11699, 11701	10/26, 2/23, 5/28, 3/29	
180	?	?	10 T open 16 ft over headstocks, 9 ft wheelbase	11657	9/35	
206	1904	Barry Co	10 T open 15 ft over headstocks, 9 ft wheelbase	11704	4/35	
216	1898	Barry Co	10 T open 16 ft over headstocks, 9 ft wheelbase	11721	8/29	Dwg BRW2 Note 2
219-238	1918	Barry Co	10 T timber sets 15 ft over headstocks, 7 ft wheelbase	8487-96, 8498-8503, 8508, 32631, 32648, 32559	Between 6/24 and 6/31	
289-292	1896	Ashbury	10 T rail sets, 16 ft over headsatocks, 9 ft wheelbase	32599, 32600, 32603-4	First 12/32, others probably after 1947	
293-302	1898	Harrison & Camm	10 T timber sets, 15 ft over headstocks, 7 ft wheelbase	32551, 32552, 32555, 32556, 32561-68	Between 2/33 and 6/39	Dwg. BRW3 Note 4
303, 304	1898	Harrison & Camm	10 T rail set, 15 ft over headstocks, 7 ft wheelbase	32605, 32606	5/36	
305-322	1898	Harrison & Camm	10 T timber sets, 15 ft over headstocks, 7 ft wheelbase	32569, 32570, 32573, 32574, 32579-86, 32591-96	Between 9/34 and 1/39	
323-342	1898	Harrison & Camm	10 T low sided 16 ft over headstocks, 9 ft wheelbase	11819, 11820, 11823, 11825-28, 11831, 11832, 11834-37, 11841, 11842, 11844, 11845, 11848, 11849, 11851	Between 3/27 and 4/39	

Barry Railway history				Subsequent history		Notes
BR No.	Built	Builder	Description	Renumbered	Withdrawn	
343-347, 350-353, 355-361, 363, 364, 367, 368, 371-73, 375-381, 385, 386, 388-90, 393-96, 398, 399, 410, 413, 415-22, 424-426, 428-32, 436-42, 444, 445, 449, 451-53, 455-61, 465, 466, 469-79, 481-86, 489, 491, 493, 495-99, 514-19, 521, 523-26, 528-32, 534, 536-41, 543, 544, 546-49, 551-59, 561-65, 567, 569-73, 575, 576, 579-83, 585, 586	1898	Harrison & Camm	10 T open 16 ft over headstocks, 9 ft wheelbase	11852, 11853, 12056, 12074, 12246, 12316, 12344, 12346, 12356, 12457, 12469, 12477, 12482, 12754, 12787, 12846, 12856, 13130, 13224, 13434, 13450, 13453, 15007, 15019, 15020, 15021, 15052, 15053, 15060, 15093, 15100, 15128, 15138, 15153, 15197, 15205, 15210, 15215, 15228, 15242, 15244, 15299, 15328, 15342, 15345, 15348, 15350, 15353, 15356, 15373, 15383, 15387, 15389, 15403, 15420, 15424, 15429, 15447, 15534, 15573, 15582, 15599, 15643, 15677, 15685, 15732, 15744, 15753, 15763, 15791, 15796, 15536, 15543, 15796, 15827, 15828, 15837, 15839, 15899, 15900, 15930, 15939, 15969, 15981, 15984, 16000, 17478, 17479, 17481-83, 17485-90, 17493, 17495, 17497, 17500, 18085, 18115, 18150, 18169, 18228, 18253, 18283, 18300, 18531, 19550, 18624, 18946, 18997, 19093, 19151, 19283, 19300, 19301, 19305, 19307, 19319, 19397, 19405, 19416, 19422, 19683, 19765, 20020, 20036, 20072, 20086, 20108, 20109, 20164, 20165, 20167, 20173, 20178, 20181, 20203, 20207, 20213, 20231, 20233, 20240, 20245, 20256, 20261, 20281, 20290, 20330, 20377, 20378, 20432, 20436, 20459, 20465, 20469, 20471, 20472, 20489, 20490	Between 1/23 and 12/41	
370	1907	Barry Co	10 T open 16 ft over headstocks, 9 ft wheelbase	13278	12/28	Dwg BRW2
400-409	1896	Brown, Marshalls & Co	10 T van 16 ft over headstocks, 9 ft wheelbase	2165, 2169, 2175, 2240, 3093, 3113, 3153, 3182, 3287, 3303	Between 8/25 and 6/35	Dwg. BRW4
500, 501	1896	Brown, Marshalls & Co	10 T cattle 16 ft over headstocks, 9 ft wheelbase	13601, 13603	10/33, 2/28	Dwg. BRW5
502-513	1897	Midland	10 T sleeper 19 ft 8 in. over headstocks, 10 ft 6 in. wheelbase	14012, 14402, 14407, 14408, 14412, 14422, 14426, 14429, 14433-36	Between 12/22 and 1/36	

Low sided wagon No. 331, built by Harrison & Camm in 1898. Channel iron underframe, Ibbotson's D4 buffers. Photographed at Barmouth in 1924. HMRS AAJ918

Barry Railway history				Subsequent history		Notes
BR No.	**Built**	**Builder**	**Description**	**Renumbered**	**Withdrawn**	
642	1894	Barry Co	10 T stores 15 ft 6 in. over headstocks, 9 ft wheelbase	101549	2/30	
643	1900	Harrison & Camm	20 T wheel & propellor 25 ft over headstocks, 16 ft wheelbase	41994	NR, possibly after 1947	
644-655	1902	Gloucester	10 T van 16 ft over headstocks, 9 ft wheelbase	5999, 8082-84, 8089, 8094, 8103, 8124, 8149, 8216, 8224, 8231	Between 5/28 and 4/38	
656-755	1904	Turner	10 T open goods wagons 16 ft over headstocks, 9 ft wheelbase	24194, 24227, 24321, 24428, 24462, 24532, 24538, 24580, 24611, 24622, 24643, 24685, 24798, 24823, 24886, 24904, 24919, 24920, 24923, 24930, 24932, 24933, 24935, 24941, 24944, 24963, 24968-70, 24977, 24983, 24996, 25102, 25105, 25106, 25109, 25113, 25128, 25135, 25145, 25162, 25168, 25169, 25177, 25178, 25181, 25187, 25188, 25197, 25204, 25216, 25223, 25231, 25235, 25238, 25243, 25247, 25249, 25252, 25253, 25257, 25273, 25275, 25279, 25290, 25291, 25297, 25300, 25304, 25311, 25315, 25320, 25331, 25346, 25350, 25353, 25356-58, 25382, 25383, 25387, 25389, 25396, 25399, 25407, 25412, 25418, 25420, 25431, 25436, 25450, 25452, 25457, 25459, 25465, 25466, 25468, 25475, 25476	First 04/23, most before 1947	
756-785	1904	Cravens	10 T steel van 16 ft over headstocks, 9 ft wheelbase	8239, 8432, 8887, 8889, 8891, 8896, 8920, 8930, 8933, 8946, 8959, 8968, 8969, 8977, 8983, 11024, 11029, 11056, 11062, 11075, 11092, 11121, 11176, 11178, 11198, 11251, 11340, 11358, 11389, 11453	First 10/22, most after 1947	Dwg. BRW6 Note 5
786	1904	Cravens	10 T special cattle 18 ft 6 in. over headstocks, 11 ft wheelbase, vacuum fitted	13604 but then 948 in Brown Vehicle series	NR	
787-792	1904	Cravens	10 T cattle 18 ft 6 in. over headstocks, 11 ft wheelbase	13605, 13609, 13610, 13612, 13613, 13614	From 12/33, most NR, possibly after 1947	

The number looks very much like 516, but the wagon is actually one of 12 sleeper wagons numbered between 502 and 513 and built in 1897 by the Midland Wagon Co and so the number is almost certainly 510. They were 19 ft 8 in. over headstocks with a wheelbase of 10 ft 6 in. As can be seen, the ends were slightly raised. They were withdrawn between 1922 and 1936. C W Harris

Barry Railway history				Subsequent history		Notes
BR No.	Built	Builder	Description	Renumbered	Withdrawn	
793-942	1905	G R Turner	10 T steel van 16 ft over headstocks, 9 ft wheelbase	11937, 11985, 11999, 12540, 12551, 12561, 12590, 16423, 16511, 16677, 16689, 16705, 16877, 16915, 16943, 16973, 22339, 22340, 22543, 22547, 22553, 22554, 22557, 22558, 22560, 22561, 22579, 22582, 22586-88, 22592, 22596, 22599, 22602-04, 22606, 22608-10, 22613-15, 22617-19, 22623-25, 22641, 27656, 27658, 27663, 27665, 27667, 27668, 27670-03, 27675, 27680, 27694, 27695, 27757, 28647-50, 28771, 28861, 28886, 35192, 35204, 35206, 35213, 35216, 35223, 35225, 35225(?), 35234, 35241, 35244, 35246, 35247, 35250, 35253, 35254, 35258, 35261, 35263, 35270, 35272, 35273, 35279, 35285, 35295, 35297, 35300, 35306, 35307, 35309, 35312, 35316, 35321, 35323-25, 35327, 35331, 35339, 35345, 35351, 35360, 35383, 35393, 35397-99, 35405-07, 35410, 35417, 25421, 35424, 35426, 35437, 35442, 35445, 35449, 35456, 35460, 35464, 35467, 35470, 35472, 35475, 35477, 35482-84, 35491, 35498, 35509-11, 35513, 35515	First 7/21, most NR, possibly after 1947	Dwg, BRW6 Note 5
943-1092	1912	Midland	10 T steel van 16 ft over headstocks, 9ft wheelbase	35517, 35519, 35522, 35524, 35525, 35529, 35530, 35533, 35546, 35571, 35574, 35576, 35580, 35581, 35590, 35595, 37005, 37008, 37009, 37011, 37014, 37017, 37019, 37020, 37022, 37025, 37027, 37034, 37038, 37041, 37042, 37046-49, 37055, 37060, 37067-69, 37072, 37073, 37078, 37081, 37083, 37084, 37090, 37091, 37095-97, 37099, 37101, 37103, 37116, 37123, 37128, 37129, 37133, 37134, 37136, 37148, 37152-56, 37158, 37160, 37167, 37169, 37174, 37177, 37180, 37183, 37185-87, 37189, 37194, 38198, 37199, 37202, 37204, 37206, 37207, 37213, 37217-19, 37227, 37234, 37238, 37242, 37249, 37254, 37258, 37260, 37264, 37266, 37268, 37271, 37279, 37982, 37288, 37290, 37292, 37293, 37295, 37299, 37301-03, 37305, 37310, 37311, 37313-17, 37320, 37324, 37332, 37335, 37336, 37338-42, 37344, 37350, 37352, 37359, 37361, 37373, 37377, 37378, 37381, 37384, 37391-93, 37400, 37402, 37406, 37409, 37415, 37418	First 12/35, most NR, possibly after 1947	

Open wagon No. 684, one of a batch of 100 built by G R Turner in 1904. Pressed steel underframe and laminated buffers. Withdrawn in November 1933. C W Harris

Barry Railway history				Subsequent history		Notes
BR No.	Built	Builder	Description	Renumbered	Withdrawn	
1143-1342	1913/4	S J Claye	10 T steel van 16 ft over headstocks, 9 ft wheelbase	37424, 37425, 37427, 37432, 37436, 37440, 37443, 37445, 37449, 37453, 37456, 37457, 37461, 37466, 37467, 37471, 37476, 37478-80, 37482, 37487, 37490-02, 37504, 37505, 37509, 37510, 37512, 37514, 37524, 37525, 37530, 37534, 37537, 37541, 37544, 37546, 37547, 37552, 37554, 37555, 37560, 37564, 37566, 37568, 37570, 37571, 37573, 37575, 37576, 37584, 37591, 37592, 37594, 37600, 37602, 37603, 37605, 47418, 47663, 47738-50, 69856, 93840-963	First 12/35, most NR, possibly after 1947 Note 8	Dwg BRW6 Note 7
1343-1542	1914	Metropolitan	10 T steel van 16 ft over headstocks, 9 ft wheelbase	93964-94000, 100500-100662	One 12/35, the rest NR, probably after 1947 Note 9	Dwg BRW6 Note 6

Notes

1. Listed by the Barry Railway in Drawing No. 49-31 as for No. 349, and therefore probably almost identical.
2. Listed by the Barry Railway in Drawing No. 74-21 as for No. 11, and therefore probably almost identical except for D4 buffers instead of laminated spring buffers with 12 plates.
3. Listed by the Barry Railway in Drawing No. 49-31 as for No. 349, and therefore probably almost identical except for having spring buffers with 14 plates.
4. Listed by the Barry Railway in Drawing No. 125-21 as for pair No. 151, and therefore probably almost identical.
5. Fox's patent pressed steel underframes
6. Channel steel underframes. Given Drawing No. 18-31 by the Barry Railway; differences from the earlier Cravens built vehicles are shown in the drawing in "All about Iron Minks" (HMRS, 1980).
7. Channel steel underframes. These vehicles are not mentioned in "All about Iron Minks" (HMRS, 1980) but were given Drawing No. 18-31 by the Barry Railway, as for the Midland and Metropolitan built vehicles, so were probably very similar.
8. Preserved, at Kent & East Sussex Railway, awaiting restoration.
9. Preserved and restored, at Gloucestershire & Warwickshire Railway.

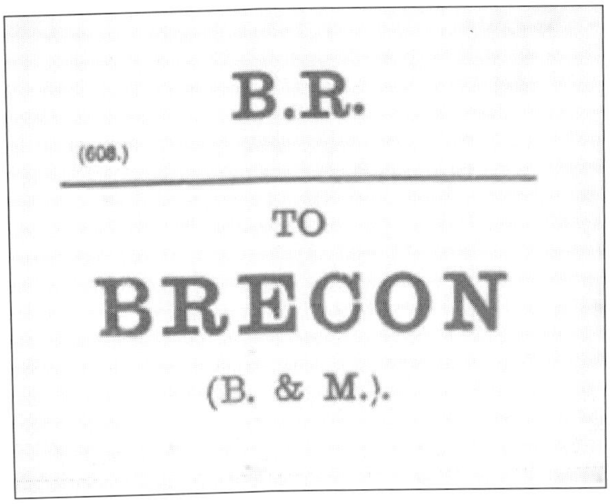

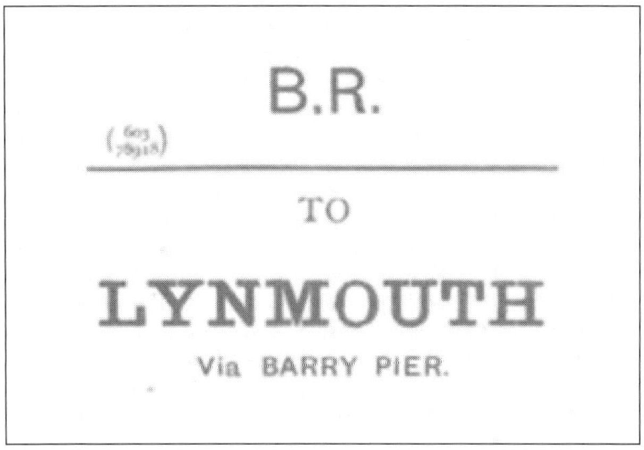

The Carrier Act of 1830 which was responsible for railways being regarded as a common carrier meant that luggage labels for all destinations around the UK network abounded; the Barry Railway was no exception. Two examples of these can be seen above. What is less well known is that the Red Funnel fleet of the Barry Railway, sailing from Barry Pier, was used for exactly the same purpose, as the North Devon destination on the label on the right shows. The main intention of the steamship operation was for passengers and excursion traffic. However the opportunity for other income streams was not lost on the Company.

Ballast Wagons

Barry Railway history				Subsequent history		Notes
BR No.	Built	Builder	Description	Renumbered	Withdrawn	
1, 2, 6-9, 14, 15, 21-35, 37-40, 44-46, 49, 52, 56, 61, 62	?	?	8 T ballast wagons, oak frames, dead buffers	14051, 14054, 14080, 14082, 14085, 14096, 14137, 14167, 14192, 14198, 14210, 14211, 14218, 14237, 14246, 14250, 14251, 14252, 14258, 14278, 14293, 14298, 14305, 14312, 14314, 14326, 14327, 14341, 14348, 14352, 14375, 14397, 14471, 14478, 14479	5/27 to 1/35	Note 1
3-5, 10, 36, 41-43, 47, 48	1889	Swansea	8 T ballast wagons, dead buffers	14060, 14064, 14073, 14108, 14308, 14230, 14231, 14234, 14356, 14368	6/27 to 1/35	Note 2
11, 12, 16-18, 20, 50, 53-55, 57-60, 63, 65, 67	?	?	8 ton ballast wagons, oak frames, Ibbotson's B4 buffers	14113, 14115, 14168-70, 14179, 14384, 14466-68, 14472-74, 14476, 14480, 14482, 14484	12/22-10/33	Note 1
13, 19, 51	1890	Ashbury	10 T ballast wagons, oak frames, B4 buffers	14126, 14171, 14396	9/23-10/27	Note 1
64	?	?	8 T ballast wagon, pressed steel underframe, Ibbotson's buffers	14481	5/29	
66	1892	Ashbury	10 T ballast wagons, oak frames, B4 buffers	14483	8/30	Note 1
68-78, 118, 120-156	1896	Ashbury	10 T ballast wagons, channel frames, sprung buffers	14485, 14486, 14488-90, 14506, 14512, 14520, 14522, 14529, 14541, 14603, 14614, 14617, 14619-29, 14633-38, 14640-45, 14647-49, 14651-59	11/22-7/33	
79-95, 157-175, 177-186, 286, 327-332	1898	Harrison & Camm	10 T ballast wagons, channel frames, sprung buffers	14546, 14548, 14553, 14560, 14567-79, 14660-14667, 14669, 14670, 14673, 14675-77, 14740-44, 14746-48, 14751, 14759-65, 14766, 14767-72	12/22-10/43	Note 3
96-100, 102, 104, 108-110	1890	Swansea	10 T ballast wagons, channel iron frames, sprung buffers	14580-85, 14587, 14589, 14593-95	4/26-10/34	Note 4
111	?	?	10 T ballast wagons, oak frames, sprung buffers	14596	9/29	Note 5

Ballast wagon No. 85, built by Harrison & Camm in 1898. Allocated GWR number 14569 but not renumbered, and withdrawn in December 1922 but possibly then transferred to Docks use as the photograph was reportedly taken in the 1920s and it carries a cross on the door. Great Western Railway negative C1/35

Barry Railway history				Subsequent history		
BR No.	Built	Builder	Description	Renumbered	Withdrawn	
112	1902	Barry Co	10 T ballast wagons, pressed steel frames, D4 buffers	14597	3/28	Note 6
113	1901	Barry Co	10 T ballast wagons, pressed steel frames, D4 buffers	14598	10/29	Note 6
115	1897	Ashbury	10 T ballast wagons, channel frames, sprung buffers	14600	10/43	Note 7
116	1892	Ashbury	10 T ballast wagons, channel frames, sprung buffers	14601	7/33	Note 7
119	?	?	8 T ballast wagon, oak frames, self contained buffers	14611	5/32	Note 8
101, 103, 105-107, 114, 117	?	?	8 T ballast wagons, oak frames, dumb and self-contained buffers	14586, 14588, 14590-92, 14599, 14602	5/26-11/32	Note 6
176	1903	Barry Co	10 T ballast wagons, pressed steel frames, D4 buffers	14745	3/28	
187-276	1889	Swansea	Side tipping ballast wagons, elm frames, dumb buffers		N/A	Note 9
277-285, 287-326	1897	Harrison & Camm	Side tipping ballast wagons, elm frames, dumb buffers		N/A	Note 10
333-342, 344-363, 365-371	?	?	10 T ballast wagons, oak frames, sprung buffers	14773, 14776, 14779, 14839, 14871, 14898-902, 14906-20, 14922-25, 40235, 40312, 40335, 40503, 40617, 40636, 40682, 40848, 40888	12/22-5/35	Note 10
343, 364	?	?	10 T ballast wagons, oak frames, sprung buffers	14903, 40316	3/28, 5/29	Note 1

Notes

1. Full length doors
2. Drop sides
3. Many to Docks Dept, some withdrawn after Nationalisation
4. Some withdrawn after Nationalisation
5. Full length doors, to Docks Dept 1/27
6. To Docks Dept
7. To Docks Dept 4/27
8. To Docks Dept 4/26
9. Allocated numbers in the series 201-339 but appear to have been sold
10. Dropside, some withdrawn after Nationalisation

Coal Wagons

Barry Railway history				Subsequent history		Notes
BR No.	Built	Builder	Description	Renumbered	Withdrawn	
1, 2, 4, 8-21, 28-30, 32-34, 38, 41, 50, 54-56, 65, 67, 516	?	?	10 T coal wagons, side doors only	9930, 9940, 9999, 23068, 23094, 23111, 23114, 23116, 23128, 23131, 23135, 23139-41, 23146, 23148, 23154, 23176, 23182, 23185, 23187, 32190, 23195, 23230, 23249, 23335, 23361, 23362, 23365, 23383, 23385, NR	From 10/22 to after Nationalisation	
3, 7, 25-27, 39	1907	Barry Co	12 T coal wagons, laminated spring buffers	9990, 23048, 23171, 23173, 23174, 23232	3/32-5/40	
5, 6	?	?	?	23010, 23039	5/40, 9/28	No information
37	?	?	10 T coal wagons, side and bottom doors	23226	7/32	
42-49, 51-53, 57-62, 64, 68-74, 95-106	?	?	10 T coal wagons, side and end doors	23254, 23257, 23271, 23284, 23315, 23321, 23327, 23328, 23332, 23336, 23349, 23359, 23370, 23371, 23374, 23378-80, 23382, 23386-92, 23413-24	From 7/32 to after Nationalisation	
22-24, 31, 35, 36, 40	?	?	10 T coal wagons, side, end and bottom doors	23157, 23163, 23170, 23186, 23198, 23201, 23243	From 5/35 to after Nationalisation	
63, 66	1903	Barry Co	12 T coal wagons, laminated spring buffers	23381, 23384	5/29, 11/26	
75-94	1900	Harrison & Camm	10 T coal wagons, Ibbotson's buffers	23393-412	From 3/29 to after Nationalisation	
107-146	1914	Pickering	10 T coal wagons, laminated spring buffers	23425-29, 23431-34, 23436, 23440, 23443, 23445, 23447, 23450-75	6/27-4/35	

Seven plank locomotive coal wagon at Penmaenpool in 1924 alongside a Cambrian Railways four plank wagon, also branded for loco coal only. This is probably one of the batch of 20 built by Harrison & Camm in 1900 and numbered 75 to 94.

HMRS Jim Richards Collection AAJ919

Miscellaneous Vehicles

Barry Railway history				Subsequent history		Notes
BR No.	Built	Builder	Description	Renumbered	Withdrawn	
151	1894	Birmingham	5 T steam crane protector for crane No. 1			
152	1895	J Jessop & Sons	5 T steam crane protector for crane No. 2			
153, 154	1898	J Jessop & Sons	5 T steam crane protector for cranes Nos. 3 and 4			
155	?	Swansea	Breakdown van			
156	1900	Barry Co	Breakdown wagon			
247, 248	1896	Ashbury	Horseboxes			
251	1896	Ashbury	Carriage truck			
379	?	?	Brake van	14027	8/36	
383	?	?	Open wagon	15536	4/31	
602	?	?	Open wagon	15543	5/37	
1	1890	Metropolitan	Pay Clerk's Van			Note
2	?	?	Inspection Carriage	80969	NR	
3	?	?	Engineers' coach	80966	1/31	
22	?	?	Engineers' coach	80967	9/28	
52	?	?	Ballast wagon	14349	10/28	
1	1899	Cravens	Bogie saloon	In passenger stock list		
1	?	?	Mess and tool van	43899	11/29	
2	?	?	Mess and tool van	43900	2/28	
Un-numbered	?	?	Pooley's weighing machine van	37001	5/30	

Note: Formerly Mineral Brake van No. 31. In 1905 it was fitted with a petrol engine but this was soon removed

Miscellaneous Duplicate Numbered Vehicles

Nos.	Description	Built	Builder	Withdrawn	Comments
03, 05, 06, 019, 020	10 T coal wagon, side doors only	?	?	?	Surplus stock
07, 021, 025, 026, 027, 030, 032, 033, 034, 035, 038, 039	10 T coal wagon, side, end and bottom doors	?	?	?	Surplus stock

Cydnabyddiaethau

Cynhyrchwyd y llyfr hwn gan gyd-ymdrechion aelodau o Weithgor Cyhoeddi Cylch Ymchwil Rheilffyrdd Cymru ac aelodau o'r Cylch, i gyd yn benderfynol o adeiladu ar lwyddiant cyfrol gyntaf Nigel Nicholson ar Reilffordd Rhymni. Yn arbennig, hoffwn ddiolch i Jonathan David, nid yn unig ar gyfer ei waith yn cysodi'r llyfr ond am ei ddiwydrwydd wrth roi mynegeion y locomotifau, cherbydau a wageni at ei gilydd. Diolch hefyd i Archifydd y Cylch, Tony Miller, am ddod o hyd i ddogfennau a ffotograffau yn fwy cloi nag y gallem fod wedi ei wneud ac, felly, yn arbed llawer o amser i ni wrth gynhyrchu'r llyfr hwn.

Acknowledgements

This book has been produced by the combined efforts of the members of the Publications Working Group of the Welsh Railways Research Circle and members of the Circle, determined to build on the success of Nigel Nicholson's first volume on the Rhymney Railway. In particular, I would like to thank Jonathan David, not just for his work in typesetting the book but for his diligence in putting together the indices of the rolling stock. Thanks must also go to the Circle's Archivist, Tony Miller, for locating important documents and photographs more quickly than we could have done and thus saving us a great deal of time in producing this book.

Llyfryddiaeth a Darllen Pellach

Bibliography and Further Reading

Ahrons, E L, Locomotive and Train Working in the Latter Part of the Nineteenth Century - Volume Four, pp 110-119 (W. Heffer & Sons Ltd, 1953)
Allcock, N J *et al*, The Locomotives of the Great Western Railway - Part Ten - Absorbed Engines 1922-1947 pp K23-K52 (The Railway Correspondence and Travel Society, 1966)
Allcock, N J *et al*, The Locomotives of the Great Western Railway - Part Eleven - The Rail Motor Vehicles and Internal Combustion Locomotives p L12 (The Railway Correspondence and Travel Society, 1952)
Barrie, D S M, The Barry Railway (The Oakwood Press, 1962, reprinted 1978)
Brown, M, Barry No. 1 Dock: The building of a giant. *Archive*, Issue 40, pp 31-47 (Lightmoor Press, 2003)
Chapman, C, The Vale of Glamorgan Railway (The Oakwood Press, 1998)
Lewis, J. Great Western Steam Rail Motors and Their Services, pp 230-231 (Wild Swan Publications Ltd, 2004)
Miller, B J, Rails to Prosperity - The Barry & After 1884 to 1984 (Regional Publications [Bristol] Limited, 1984)
Morgan, H, South Wales Branch Lines, pp 14-25 (Ian Allan Ltd, 1984)
Mountford, E R, A Register of G.W.R. Absorbed Coaching Stock 1922/3, pp 33-39 (The Oakwood Press, 1978)
Mountford, E R, The Barry Railway Diagrams and photographs of Locomotives, Coaches and Wagons (The Oakwood Press, 1987)
Rush, R W, Lancashire & Yorkshire Passenger Stock, pp 63-66 (The Oakwood Press, 1984)
Slinn, J N, Great Western Way, pp 189-196 (Historical Model Railway Society, 1978)

This photograph poses a slight problem. It is claimed to be of wagon No. 52. However, the only timber pairs were numbered 61 to 66, 141 to 154, 219 to 238 and 298 to 322. The latter were 10 ft wheelbase and Nos. 219 to 238 had timber underframes, whereas these seem to be 7 ft wheelbase with channel underframes. It is therefore probably No. 152. C W Harris